Worksheets
For Classroom or Lab Practice

Patricia Foard
SOUTH PLAINS COLLEGE

Prealgebra

FIFTH EDITION

Elayn Martin-Gay

PEARSON

Prentice Hall

Upper Saddle River, NJ 07458

Vice President and Editorial Director, Mathematics: Christine Hoag
Executive Editor: Paul Murphy
Project Manager, Editorial: Mary Beckwith
Editorial Assistant: Georgina Brown
Senior Managing Editor: Linda Mihatov Behrens
Associate Managing Editor: Bayani Mendoza de Leon
Project Manager, Production: Barbara Mack
Supplement Cover Manager: Paul Gourhan
Supplement Cover Designer: Victoria Colotta
Operations Specialist: Ilene Kahn
Senior Operations Supervisor: Diane Peirano
Cover Photo: Getty Images

© 2008 Pearson Education, Inc.
Pearson Prentice Hall
Pearson Education, Inc.
Upper Saddle River, NJ 07458

The author and publisher of this book have used their best efforts in preparing this book. These efforts include the development, research, and testing of the theories and programs to determine their effectiveness. The author and publisher make no warranty of any kind, expressed or implied, with regard to these programs or the documentation contained in this book. The author and publisher shall not be liable in any event for incidental or consequential damages in connection with, or arising out of, the furnishing, performance, or use of these programs.

Printed in the United States of America

10 9 8 7 6 5 4 3 2 1

ISBN 13: 978-0-13-235400-4 Standalone

ISBN 10:   0-13-235400-4 Standalone

ISBN 13: 978-0-13-600860-6 Value Pack

ISBN 10:   0-13-600860-7 Value Pack

Pearson Education Ltd., *London*
Pearson Education Australia Pty. Ltd., *Sydney*
Pearson Education Singapore, Pte. Ltd.
Pearson Education North Asia Ltd., *Hong Kong*
Pearson Education Canada, Inc., *Toronto*
Pearson Educación de Mexico, S.A. de C.V.
Pearson Education—Japan, *Tokyo*
Pearson Education Malaysia, Pte. Ltd.

Worksheets
Prealgebra, Fifth Edition
Table of Contents

1.2 Place Value and Names for Numbers

Learning Objectives
    A. Find the place value of a digit in a whole number.
    B. Write a whole number in words and in standard form.
    C. Write a whole number in expanded form.
    D. Read tables.

Key Vocabulary
digits, whole numbers, place value, standard form, period, expanded form

Objective A **Find the place value of a digit in a whole number.**

   Example 1 Find the place value of the digit 7 in each whole number.

    a) 175,128   b) 12,715   c) 17,213,489

Objective B **Write a whole number in words and in standard form.**

   Example 2 Write each number in words.

    a) 37   b) 349

    c) 113,143   d) 834,193,038

   Example 3 Write each number in standard form.

    a) seventy-eight   b) five hundred twenty-five

    c) two million, four hundred fifty-three thousand, nine

Answers: 1a) ten-thousands, b) hundreds, c) millions, 2a) thirty-seven, b) three hundred forty-nine, c) one hundred thirteen thousand, one hundred forty-three, d) eight hundred thirty-four million, one hundred ninety-three thousand, thirty-eight, 3a) 78, b) 525, c) 2,453,009,

Examples 1.2 (cont'd)

Objective C Write a whole number in expanded form.

    Example 4 Write 283,197 in expanded form.

Objective D Read tables.

    Example 5 Use the table to answer each question.

Grades in Statistical Methods					
	A	B	C	D	F or W
Freshman	17	21	28	12	10
Sophomore	20	18	35	8	12

    a) How many sophomores made an A?

    b) Which classification had more D's?

4) 200,000 + 80,000 + 3000 + 100 + 90 + 7, 5a) 20, b) freshman

Practice Set 1.2

Use the choices below to fill in each blank.

standard form   period   whole
expanded form   place value   words

1. The number fifty-eight is written in _____.

2. The number 378 is written in _____.

3. The number $1000 + 300 + 60 + 9$ is written in _____.

4. In a whole number, each group of three digits is called a _____.

5. In the number 3,789, the number 7 is in the hundreds _____.

6. The number 0, 1, 2, 3, 4, . . . are called _____ numbers.

Objective A Find the place value of a digit in a whole number.
Determine the place value of the digit 8 in each whole number.

7. 185

7. _____

8. 8,306

8. _____

9. 183,435

9. _____

Objective B Write each number in words.

10. 403

10. _____

11. 23,831

11. _____

12. 16,390,569

12. _____

13. The current population of Levelland, Texas is 13,126.

13. _____

_____

14. A clinic saw 1883 patients one month.

14. _____

_____

Practice Set 1.2 (cont'd)

Write each whole number in standard form.

15. Nine thousand, eight hundred sixteen

15. _____

16. Three million, two hundred fifty-five thousand, twelve

16. _____

Objective C Write a whole number in expanded form.
Write each whole number in expanded form.

17. 13,209

17. _____

18. 3921

18. _____

19. 12,309,141

19. _____

Objective D Read tables.
Use the table to answer the following questions.

A recent survey of 250 men and 250 women gave the following results.

	Ate a Healthy Breakfast	Exercised 2 - 3 Times a Week
Men	176	176
Women	205	135

20. Did more men or more women eat a healthy breakfast?

20. _____

21. How many women exercised 2 - 3 times a week?

21. _____

22. How many men ate a healthy breakfast?

22. _____

Extensions

23. Write the largest 4-digit number that can be made.

23. _____

24. Write the largest four-digit number that can be made if no digit can be used more than once.

24. _____

25. Check to see whether the number written in standard form matches the number written in words. If not, explain what is wrong and write the number correctly in words.

25. _____

206   two hundred and sixty

1.3 Adding and Subtracting Whole Numbers, and Perimeter

Learning Objectives
  A. Add whole numbers
  B. Subtract whole numbers.
  C. Find the perimeter of a polygon.
  D. Solve problems by adding or subtracting whole numbers.

Key Vocabulary
sum, addend, associative, commutative, perimeter, minuend, subtrahend, difference

---

Objective A Add whole numbers.

  Example 1 Add: $23 + 113$

  Example 2 Add: $2{,}398 + 12{,}943$

  Example 3 Add: $12 + 8 + 5 + 3 + 2$

  Example 4 Add: $295 + 1290 + 29 + 238$

Objective B Subtract whole numbers.

  Example 5 Subtract. Check each answer by adding.

  a) $13 - 5$ b) $25 - 6$ c) $28 - 0$ d) $17 - 17$

  Example 6 Subtract. Check by adding. $3987 - 635$

  Example 7 Subtract. Check by adding. $971 - 89$

1) 136, 2) 15,341, 3) 30, 4) 1852, 5a) 8, b) 19, c) 28, d) 0, 6) 3352, 7) 882

Examples 1.3 (cont'd)

Example 8 Subtract. Check by adding. $1000 - 539$

Objective C Find the perimeter of a polygon.

Example 9 Find the perimeter of the polygon shown.

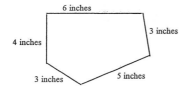

Example 10
A classroom is in the shape of a rectangle with width of 60 feet and length of 85 feet. Find the perimeter of this room.

Objective D Solve the problems by adding or subtracting whole numbers.

Example 11
Find the sum of 298 and 174.

Example 12
Find the difference of 17 and 9.

Example 13 Use the bar graph to answer the questions.

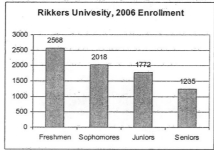

a) Which classification had the most students enrolled?

b) How many freshmen and sophomores were enrolled?

8) 461, 9) 21 inches, 10) 290 feet, 11) 472, 12) 8, 13a) freshmen, b) 4586

Practice Set 1.3

Use the choices below to fill in each blank.

addend	associative	commutative	difference
minuend	perimeter	subtrahend	sum

1. In $27 + 8 = 35$, the numbers 27 and 8 are each called a(n) _____ and the number 35 is called the _____ .

2. The distance around a polygon is called its _____ .

3. In $28 - 6 = 22$, the number 28 is called the _____ , the 6 is called the _____ , and 22 is called the _____ .

4. $12 + 8 = 8 + 12$ This property is called the _____ property of addition.

5. $(2 + 3) + 6 = 2 + (3 + 6)$ This property is called the _____ property of addition.

Objective A   Add whole numbers.
Add.

6. $12 + 15$ 6. _____

7. $\begin{array}{r} 5267 \\ +936 \\ \hline \end{array}$ 7. _____

8. $26 + 235 + 198 + 339$ 8. _____

Objective B   Subtract whole numbers.
Subtract. Check by adding.

9. $\begin{array}{r} 48 \\ -19 \\ \hline \end{array}$ 9. _____

Practice Set 1.3 (cont'd)

10. 1953 −844

10. _____

11. 29,391 − 16,492

11. _____

Objective C Find the perimeter of a polygon.
Find the perimeter of each figure.

12.

12. _____

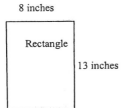

8 inches

Rectangle

13 inches

13.

13. _____

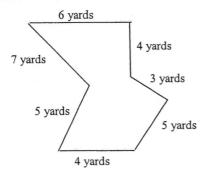

6 yards

4 yards

7 yards

3 yards

5 yards

5 yards

4 yards

Practice Set 1.3 (cont'd)

Objective D Solve the problems by adding or subtracting whole numbers.

14. Find the sum of 138 and 2048.   **14.** _____

15. Find the difference of 193 and 97.   **15.** _____

16. The average distance from Earth to the moon   **16.** _____
is 92,955,800 miles. The average distance
from Mercury to the sun is 36,252,762 miles.
Find the difference of 92,955,800 and
36,252,762.

17. A television that normally sells for $478 is   **17.** _____
discounted $89 in a sale. What is the sale
price?

Practice Set 1.3 (cont'd)

Mary kept track of her expenses for a month and made a graph of the results. Use the bar graph to answer the following questions.

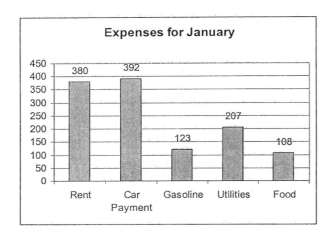

18. What was her biggest expense?   18. _____

19. How much did she spend on transportation   19. _____
       (car payment and gasoline)?

20. How much more did she spend on rent than   20. _____
       on utilities?

Extensions

21. Subtract. Identify the minuend, the   21. _____
       subtrahend, and the difference.
       184
       −95

22. Subtract 17 from 22.   22. _____

1.4 Rounding and Estimating

Learning Objectives
- A. Round whole numbers.
- B. Use rounding to estimate sums and differences.
- C. Solve problems by estimating.

Key Vocabulary
rounding, estimate, exact

Objective A Round whole numbers.

Example 1 Round 175 to the nearest ten.

Example 2 Round 37,905 to the nearest thousand.

Example 3 Round 198,349 to the nearest hundred.

Objective B Use rounding to estimate sums and difference.

Example 4 Round each number to the nearest ten to find an estimated sum.
```
 28
 36
 22
+12
```

Example 5 Round each number to the nearest hundred to find an estimate difference.
```
 2305
−1892
```

1) 180, 2) 38,000, 3) 198,300, 4) 100, 5) 400

Examples 1.4 (cont'd)

Objective C Solve problems by estimating.

Example 6 Anne wrote four checks in one day. Estimate the amount of money she spent by rounding each check she wrote to the nearest ten and adding the amounts.

$23.97   Shoes
$48.24   Coat
$137.56   Car Repair
$296.00   Rent

Example 7 A clinic in a small town reported the number of cases of flu that were treated for each month of the flu season last year: 1239, 2486, 1103, 1720. Estimate the total number of cases of flu by rounding each number to the nearest hundred.

6) $510, 7) 6500

Practice Set 1.4

Use the choices below to fill in each blank.

       estimate   **exact**   **rounding**

1. 3496 is a(n) _____ number , but 3500 would be a(n)

   _____.

2. _____ a whole number means to approximate it.

Objective A   Round whole numbers.

3. Round 397 to the nearest hundred.   3. _____

4. Round 1294 to the nearest thousand.   4. _____

5. Round 72 to the nearest ten.   5. _____

6. Round 130,952 to the nearest ten-thousand.   6. _____

Complete the table by estimating the given number to the given place value.

		Tens	Hundreds	Thousands
7.	4391			
8.	123,493			
9.	28,951			
10.	148,821			

11. A company reported that the average annual salary for its   11. _____
    executives was $253,971. Round this to the nearest ten
    thousand.

Objective B   Use rounding to estimate sums and differences.

Estimate the sum or difference by rounding each number to the nearest ten.

12. 27   12. _____
    36
    44
  +72

13. 378   13. _____
  −194

Practice Set 1.4 (cont'd)

Estimate the sum or difference by rounding each number to the nearest hundred.

14. 125
       792
      +294

14. _____

15. 3298
      −2975

15. _____

Objective C Solve problems by estimating.

16. The Anderson family took a trip and traveled 296, 562, 305, and 692 miles on four consecutive days. Round to the nearest ten and estimate the total distance they traveled.

16. _____

17. A student is pricing a new computer system. The computer costs $798, the monitor costs $326, the keyboard costs $18, and the mouse costs $26. Round each number to the nearest ten and find the total estimated cost.

17. _____

Extensions

18. Find one number that when rounded to the nearest hundred is 1200.

18. _____

19. A number rounded to the nearest hundred is 2700.
    a. Determine the smallest possible number.
    b. Determine the largest possible number .

19a. _____
19b. _____

20. Estimate the perimeter of the rectangle by first rounding each number to the nearest ten.

20. _____

42 feet

128 feet

1.5 Multiplying Whole Numbers and Area

Learning Objectives
   A. Use the properties of multiplication.
   B. Multiply whole numbers.
   C. Find the area of a rectangle.
   D. Solve problems of multiplying whole numbers.

Key Vocabulary
factor, product, distribute, area

Objective A Use the properties of multiplication.

Example 1 Multiply.

a) 8×1 b) 9×0 c) $1(19)$ d) $0 \cdot 2$

Example 2 Rewrite each using the distributive property.

a) $2(3 + 6)$ b) $19(4 + 8)$ c) $2(8 + 3)$

Objective B Multiply whole numbers.

Example 3 Multiply.

a)   16
    ×3

b)   28
    ×6

Example 4 Multiply: 342×28

Example 5 Multiply: 781×136

1a) 8, b) 0, c) 19, d) 0, 2a) 2·3 + 2·6, b) 19·4 + 19·8, c) 2·8 + 2·3, 3a) 48, b) 168, 4) 9576, 5) 106,216,

Examples 1.5 (cont'd)

Objective C Find the area of a rectangle

Example 6 A room is in the shape of a rectangle with length of 45 feet and width of 30 feet. Find its area.

Objective D Solve problems by multiplying.

Example 7 A blouse costs $28. If May buys 3 blouses, find her cost before taxes.

Example 8 Two families are going to visit a theme park. The cost for an all-day ticket is $48 for adults and $35 for children. The group has 4 adults and 6 children. What will be the total cost for the tickets?

Example 9 Each page of a book contains an average of 259 words. Estimate, rounding each number to the nearest hundred, the number of words contained on 379 pages.

6) 1350 square feet, 7) $84, 8) $402, 9) 120,000 words

Practice Set 1.5

Use the choices below to fill in each blank.

area	associative	commutative
distributive	factor	product

1. In $6 \cdot 2 = 12$, the 12 is called the _____ and 6 and 2 are called _____s.

2. When we write $3 \cdot 10 = 10 \cdot 3$, we are using the _____ property of multiplication.

3. When we write $(2 \cdot 3) \cdot 5 = 2 \cdot (3 \cdot 5)$, we are using the _____ property of multiplication.

4. The surface of a region is called the _____.

5. When we write $5(6 + 2) = 5 \cdot 6 + 5 \cdot 2$, we are using the _____ property.

Objective A Use the properties of multiplication.

Multiply.

6. $17 \cdot 1$ 6. _____

7. $1(23)$ 7. _____

8. $8 \cdot 2 \cdot 0$ 8. _____

9. $9 (0)$ 9. _____

Use the distributive property to rewrite each expression.

10. $3(5 + 8)$ 10. _____

11. $12(9 + 4)$ 11. _____

Practice Set 1.5 (cont'd)

Objective B Multiply whole numbers.

Multiply.

12. 27
    ×3

12. _____

13. 482
    ×8

13. _____

14. 129 × 28

14. _____

15. 8391
    ×138

15. _____

Objective C Find the area of a rectangle

16. Find the area of the rectangle.

16. _____

3 inches

7 inches

Practice Set 1.5 (cont'd)

17. Find the area of the rectangle.

17. _____

35 feet

25 feet

Objective D Solve problems by multiplying.

18. Multiply 82 by 17.

18. _____

19. Find the product of 9 and 138.

19. _____

20. If one ounce of cola has 20 calories, how many calories does 12 ounces of cola have?

20. _____

21. The textbook for a math course costs $112. There are 27 students in one class. Find the total cost of the books for the class.

21. _____

Practice Set 1.5 (cont'd)

22. A plot of land measures 120 feet by 97 feet. Find its area.   **22.** _____

23. Estimate the product by rounding each factor to the
nearest ten. 321×21 **23.** _____

Extensions

24. Rewrite 2×5 as repeated addition of the number two.   **24.** _____

1.6 Dividing Whole Numbers

Learning Objectives
   A. Divide whole numbers.
   B. Perform long division.
   C. Solve problems that require dividing by whole numbers.
   D. Find the average of a list of numbers.

Key Vocabulary
dividend, divisor, quotient, average

Objective A Divide whole numbers.

Example 1 Find each quotient. Check by multiplying.

a) $36 \div 9$ b) $\dfrac{72}{8}$ c) $5\overline{)35}$

Example 2 Find each quotient. Check by multiplying.

a) $1\overline{)24}$ b) $17 \div 1$ c) $\dfrac{10}{10}$ d) $19 \div 19$ e) $\dfrac{14}{1}$ f) $8\overline{)8}$

1a) 4, b) 9, c) 7, 2a) 24, b) 17, c) 1, d) 4, e) 14, f) 1

Examples 1.6 (cont'd)

Example 3 Find each quotient. Check by multiplying.

a) $0 \div 9$ b) $3\overline{)0}$ c) $\dfrac{4}{0}$ d) $\dfrac{0}{8}$

Objective B Perform long division.

Example 4 Divide: $2176 \div 8$. Check by multiplying.

Example 5 Divide and check: $294 \div 6$

3a) 0, b) 0, c) undefined, d) 0, 4) 272, 5) 49

Examples 1.6 (cont'd)

Example 6 Divide and check: $907 \div 8$

Example 7 Divide and check. $12,108 \div 7$

Example 8 Divide and check. $3198 \div 15$

6) 113 R 3, 7) 1729 R 5, 8) 213 R 3

Examples 1.6 (cont'd)

Example 9 Divide: $6550 \div 13$

Objective C Solve problems by dividing.

Example 10 Four college share an apartment. The total rent is $660. How much does each student pay?

Example 11 A club has 585 raffle tickets to sell. There are 13 members in the club. How many tickets does each member need to sell?

Objective D Find the average of a list of numbers.

Example 12 Jon made 69, 43, 89 and 95 on four quizzes. What is his quiz average?

9) 503 R 11, 10) $165, 11) 45 tickets, 12) 74

Name: Date:
Instructor: Section:

Practice Set 1.6

Use the choices below to fill in each blank.

dividend   divisor   quotient   undefined

1. Any number divided by zero is _____.

2. In 48 ÷ 16 = 3, the number 3 is called the _____, 48 is called the
_____ and 16 is called the _____.

Objective A   Divide whole numbers.

Find each quotient.

3. $18 \div 3$ **3.** _____

4. $45 \div 5$ **4.** _____

5. $81 \div 9$ **5.** _____

6. $0 \div 8$ **6.** _____

7. $42 \div 1$ **7.** _____

 8. $19 \div 19$ **8.** _____

9. $\dfrac{0}{12}$ **9.** _____

10. $\dfrac{2}{0}$ **10.** _____

11. $28 \div 1$ **11.** _____

Practice Set 1.6 (cont'd)

Objective B Perform long division.

12. $2\overline{)468}$

12. _____

13. $5\overline{)375}$

13. _____

14. $84 \div 6$

14. _____

15. $7\overline{)128}$

15. _____

16. $5136 \div 16$

16. _____

17. $48\overline{)2984}$

17. _____

Practice Set 1.6 (cont'd)

18. $27\overline{)3110}$

18. _____

19. $126\overline{)73,466}$

19. _____

Objective C Solve problems by dividing.

20. Find the quotient of 112 and 14.

20. _____

22. Thirty-eight people pooled their money and bought lottery tickets. One ticket won a prize of $7,030,000. Find how much each person received.

22. _____

Practice Set 1.6 (cont'd)

22. Marissa has to make long dresses for 12 students in a dance class. Each dress requires 6 yards. She finds 75 yards of material on sale. Determine whether 75 yards is enough to make 12 dresses. Determine the amount left over or the amount short.

22. _____

Objective D Find the average of a list of numbers.

23. Find the average: 12, 19, 24, 25, 28, 36

23. _____

Extensions

24. The area of a rectangle is 96 square feet. If the length is 12 feet, find the width.

24. _____

25. Write down any two numbers whose quotient is 24.

25. _____

1.7 Exponents and Order of Operations

Learning Objectives
  A. Write repeated factors using exponential notation.
  B. Evaluate expressions containing exponents.
  C. Use the order of operations.
  D. Find the area of a square.

Key Vocabulary
base, exponent

Objective A Write repeated factors using exponential notation.

Write using exponential notation.

Example 1 $5 \cdot 5 \cdot 5 \cdot 5 \cdot 5$

Example 2 $8 \cdot 8$

Example 3 $2 \cdot 4 \cdot 4 \cdot 4 \cdot 4 \cdot 4 \cdot 4 \cdot 4$

Example 4 $5 \cdot 5 \cdot 5 \cdot 7 \cdot 7 \cdot 7 \cdot 7$

Objective B Evaluate expressions containing exponents.

Evaluate.

Example 5 7^2

Example 6 8^1

Example 7 $3 \cdot 2^5$

Example 8 $2^2 \cdot 3^3$

Objective C Use the order of operations

Example 9 Simplify. $5 \cdot 3 + 8 \div 4$

1) 5^5, 2) 8^2, 3) $2 \cdot 4^7$, 4) $5^3 \cdot 7^4$, 5) 49, 6) 8, 7) 96, 8) 108, 9) 17

Examples 1.7 (cont'd)

Example 10 Simplify. $3^3 \div 9 - 3$

Example 11 Simplify. $3 + 4(8 - 2) + 3^2$

Example 12 Simplify. $2 + 5[4^2 - (9 \div 3)] - 6^2$

Example 13 ·Simplify. $\dfrac{2 + 3(5 - 3)}{4(3 - 2)}$

Example 14 Simplify. $16 \div 4 \cdot 2 + 6$

Objective D Find the area of the square.

Example 15 Find the area.

8 inches

10) 0, 11) 36, 12) 31, 13) 2, 14) 14, 15) 64 square inches

Practice Set 1.7

Use the choices below to fill in each blank.

   base   **exponent**

1. In $3^5 = 243$, the 3 is called the _____ and the 5 is called the
   _____ .

Objective A   Write repeated factors using exponential notation.

Write using exponential notation.

2. $3 \cdot 3 \cdot 3 \cdot 3 \cdot 3 \cdot 3$ 2. _____

3. $12 \cdot 12 \cdot 12 \cdot 12$ 3. _____

4. $2 \cdot 3 \cdot 3 \cdot 3 \cdot 5 \cdot 5 \cdot 5$ 4. _____

Objective B   Evaluate expressions containing exponents.

Evaluate.

5. 2^3 5. _____

6. 3^5 6. _____

7. 12^1 7. _____

8. $2 \cdot 3^3$ 8. _____

9. $3^2 \cdot 2^2$ 9. _____

Practice Set 1.7 (cont'd)

10. $3 \cdot 8^2$ 10. _____

Objective C Use the order of operations.

11. $2 + 3 \cdot 4$ 11. _____

12. $2 \cdot 5 + 6$ 12. _____

13. $24 \div 6 \cdot 2 + 3$ 13. _____

14. $81 \div 9 - 3$ 14. _____

15. $4 + \dfrac{16}{2}$ 15. _____

16. $2^2 \cdot (3 + 4)$ 16. _____

17. $5^2 (5 - 1) + 2^2$ 17. _____

18. $(2 + 3)^2 + 4 \cdot 3^2$ 18. _____

Practice Set 1.7 (cont'd)

19. $\dfrac{3+6}{9-6}$

19. _____

20. $\dfrac{2(8-4)+2}{3^2-4}$

20. _____

21. $8 \div 2 + 4^2 - 10$

21. _____

22. $\dfrac{2(2+5)}{2^2-4}$

22. _____

Practice Set 1.7 (cont'd)

Objective D Find the area of the square.

23. Find the area of the square.   23. _____

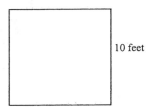

10 feet

24. Find the area of the square.   24. _____

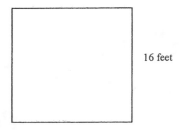

16 feet

Extensions

25. Find the perimeter of the figure.   25. _____

20 inches

6 inches

15 inches

?

?

14 inches

1.8 Introduction to Variables, Algebraic Expressions, and Equations

---

Learning Objectives
    A. Evaluate algebraic expressions given replacement values.
    B. Identify solutions of equations.
    C. Translate phrases into variable expression.

Key Vocabulary
variable, algebraic expression, equation, solution

Objective A Evaluate algebraic expressions given replacement values.

    Example 1 Evaluate $x + 3$ if x is 4.

    Example 2 Evaluate $3(x + 2y)$ for $x = 2$ and $y = 4$.

    Example 3 Evaluate $\dfrac{x - y}{x - 2y}$ for $x = 25$ and $y = 10$.

    Example 4 Evaluate $x^2 + 2x - 3$ for $x = 4$.

1) 7, 2) 30, 3) 3, 4) 21

Name: Date:
Instructor: Section:

Examples 1.8 (cont'd)

Example 5 The expression $\dfrac{9C}{5} + 32$ can be used to write degrees Celsius C as degrees Fahrenheit F. Find the value of this express for $C = 40$.

Objective B Identify solutions of equations.

Example 6 Determine whether 4 is a solution of the equation $2(x+3) = 14$.

Example 7 Determine which numbers in the set $\{10, 15, 20\}$ are solutions of the equation $3x - 5 = 40$.

Objective C Translate phrases into variable expressions.

Example 8 Write as an algebraic expression. Use x to represent "a number."

a) 8 increased by a number

b) 9 decreased by a number

c) the product of 4 and a number

d) the quotient of 6 and a number

e) a number subtracted from 8

5) 104, 6) yes, 7) 15, 8a) $8 + x$, b) $9 - x$, c) $4x$, d) $\dfrac{6}{x}$, e) $8 - x$

Martin-Gay **Prealgebra** edition 5   36

Practice Set 1.8

Use the choices below to fill in each blank. Some words may used more than once.

     equation   **expression**   **solution**   **variable**

1. A(n) _____ is a letter that represents a number.

2. An _____ contains an equal and an _____ does not.

3. In the expression $3x + 5$, x is called a(n) _____.

4. A value that replaces the variable and makes the two sides of the equation equal is called a(n) _____ to the equation.

Objective A Evaluate algebraic expressions given replacement values.

Evaluate each of the following expressions for $x = 4$, $y = 3$, and $z = 5$.

5. $2 + 3x$ **5.** _____

6. $2x^2 + yz$ **6.** _____

7. $2 + x(x + z)$ **7.** _____

8. $\dfrac{4 + 2x}{4y}$ **8.** _____

9. $\dfrac{4y}{x} + \dfrac{2z}{5}$ **9.** _____

10. $(2x + 3)^2$ **10.** _____

Practice Set 1.8 (cont'd)

11. $2x^2 + 9$ 11. _____

12. $3y(x + 2z)$ 12. _____

13. $(3z - xy)^4$ 13. _____

Objective B Identify solutions of equations.

14. Is 3 a solution of $2x - 3 = 3$?   14. _____

Objective C Translate phrases into variable expressions.

15. the sum of 4 and a number   15. _____

16. the difference of 8 and a number   16. _____

17. 7 less than a number   17. _____

18. the product of 6 and a number   18. _____

19. 5 subtracted from a number   19. _____

Extensions

20. Are the following two statements the same. If not, explain   20. _____
the difference.
7 less than a number   7 less a number

Chapter 1 Vocabulary Reference Sheet

Term	Definition	Example
Section 1.2		
Digit	Can be used to write numbers	0, 1, 2, 3, 4, 5, 6, 7, 8, 9
Whole number	0, 1, 2, 3, 4, . . .	279 is a whole number
Place value	The position of each digit in a whole number.	The digit 6 in 468 is in the ten place value.
Standard form	A whole number written with digits.	2,398,013 is in standard form.
Period	Each group of theer digits separated by a comma in a whole number.	524,123,095 524 is a period.
Expanded form	Each digit in a whole number written with its place value.	$500 + 60 + 2$ is written in expanded form.
Section 1.3		
Sum	The result of an addition problem.	$2 + 4 = 6$, 6 is the sum
Addend	Each value to be added in an addition problem	$2 + 4 = 6$, 2 and 4 are addends
Commutative	Addition is commutative since numbers can be added in any order.	$3 + 6 = 6 + 3$
Associative	Addition is associative since the grouping of numbers can be changed.	$(2 + 4) + 5 = 2 + (4 + 5)$
Perimeter	The distance around a polygon.	The distance around a triangle with sides 3, 5 and 9 would be 17.
Minuend	The number that has another value subtracted from it in a subtraction problem.	$5 - 3 = 2$, 5 is the minuend.
Subtrahend	The value that is subtracted in a subtraction problem.	$5 - 3 = 2$, 3 is the subtrahend.
Difference	The result of a subtraction problem.	$5 - 3 = 2$, 2 is the difference.
Section 1.4		
Rounding	Approximating a whole number.	279 rounded to the ten place is 280.
Estimate	The result of an operation on rounded values.	For $27 + 38$, the estimate found by rounding to the ten place is $30 + 40 = 70$.
Exact	A value that has not been rounded.	249 is exact.
Section 1.5		
Factor	Any value that is multiplied.	$2 \times 8 = 16$, 2 and 8 are factors.
Product	The result of multiplication.	$2 \times 8 = 16$, 16 is the product.
Distribute	Multiplication distributes over addition.	$5(3 + 2) = 5 \cdot 3 + 5 \cdot 2$
Area	The surface of a region.	For a rectangle with sides 4 inches and 5 inches, the area is 20 square inches.
Section 1.6		
Dividend	A value that is divided by another value.	$8 \div 2 = 4$, 8 is the dividend.
Divisor	A value that is divided into another value.	$8 \div 2 = 4$, 2 is the divisor.
Quotient	The result of division.	$8 \div 2 = 4$, 4 is the quotient.
Undefined	Division by 0 is undefined.	$12 \div 0$ is undefined.
Section 1.7		
Base	A number to be used in repeated multiplications.	5^3, 5 is the base.
Exponent	Indicates the number of times a base is to be used in repeated multiplications.	5^3, the exponent 3 indicates that 5 is to be multiplied 3 times.
Section 1.8		
Variable	A letter used to replace a number.	$2x + 3$, x is a variable.
Algebraic Expression	A combination of operations on variables and numbers.	$2x^2 + 3x - 3$ is an expression.
Equation	Expression = expression	$2x + 4 = 3x + 6$ is an equation.
Solution	A value for a variable that makes an equation a true statement.	$2 + x = 5$, 3 is a solution.

NOTES:

Chapter 1 Practice Test A

1. Write 876,124 in words.   1. _____

2. Write "seventy-two thousand eight" in standard form.   2. _____

Simplify.
3. $157 + 69$ 3. _____

4. $290 - 185$ 4. _____

5. 372×25 5. _____

6. $4681 \div 35$ 6. _____

7. $3^3 \cdot 4^2$ 7. _____

8. $15 \div 1$ 8. _____

9. $0 \div 8$ 9. _____

10. $18 \div 0$ 10. _____

11. $(4-2)^2 - 3$ 11. _____

Chapter 1 Practice Test A (cont'd)

12. $16 + 2 \cdot 5 - 4$ 12. _____

13. $5 \cdot 4^2$ 13. _____

14. $3 + 4[2 + (8 - 3)] + 2^2$ 14. _____

15. $\dfrac{8 + 4}{3 \cdot 4}$ 15. _____

16. Find the average of 32, 47, 63, 70, and 83.   16. _____

17. Round 157,298 to the nearest ten thousand.   17. _____

18. Round each number to the nearest hundred and   18. _____
     estimate the sum. 5608 + 2916 + 3290

Solve.

19. 8 less than 12   19. _____

20. Find the sum of 12 and 47.   20. _____

Chapter 1 Practice Test A (cont'd)

21. Find the product of 12 and 7.   21. _____

22. Find the quotient of 204 and 12.   22. _____

23. Thirty-one printer cartridges cost $558. How much did   23. _____
each cartridge cost?

24. Rudy looked at two mountain bikes. One cost $629 and the   24. _____
other cost $594. How much more expensive was the higher
priced bike?

25. One case of oil costs $28. What would 7 cases cost?   25. _____

26. A school ordered 12 boxes of paper that cost $26 each and   26. _____
7 calculators that cost $85 each. Find the total cost.

Find the area and perimeter of each figure.

27.   27. _____

square   8 feet

Chapter 1 Practice Test A (cont'd)

28.   28. _____

12 inches

┌─────────────────────┐
│   │
│   Rectangle   │ 8 inches
│   │
└─────────────────────┘

29. Evaluate $3(x-2)^2$ for $x = 4$.   29. _____

30. Evaluate $\dfrac{2x+4}{x+2}$ for $x = 5$.   30. _____

Translate the following into a mathematical expression. Use x to represent "a number."

31. The sum of 10 and a number   31. _____

32. The product of 6 and a number   32. _____

33. Is 4 a solution to $2x + 3 = 11$?   33. _____

Chapter 1 Practice Test B

1. Write 497,108 in words.
   a. Four hundred ninety-seven thousand, eighteen
   b. Four hundred ninety-seven thousand, one hundred and eight.
   c. Four hundred ninety-seven thousand, one hundred eighty
   d. Forty-nine thousand, seven hundred eighteen

2. Write seventy-six thousand, five hundred ten in standard form.
   a. 76,510   b. 76, 501   c. 765,010   d. 76,051

Simplify.
3. $78 + 37$
   a. 105   b. 104   c. 114   d. 115

4. $72 - 72$
   a. 1   b. 0   c. 72   d. undefined

5. $612 \div 18$
   a. 32   b. 27   c. 29   d. 34

6. $72 \cdot 1$
   a. 72   b. 1   c. 73   d. 71

7. $347 - 195$
   a. 241   b. 252   c. 192   d. 152

8. $2^3 \cdot 4$
   a. 512   b. 32   c. 24   d. 12

9. $0 \div 7$
   a. 0   b. 7   c. 70   d. undefined

10. $8 \div 0$
    a. 0   b. 8   c. undefined   d. 80

Chapter 1 Practice Test B (cont'd)

11. $3 + 2(5 + 2)$
    a. 35   **b.** 27   **c.** 17   **d.** 15

12. $4 + 16 \div 4 \cdot 2$
    a. 12   **b.** 10   **c.** 6   **d.** 3

13. $3^2 \cdot 4^3$
    a. 144   **b.** 576   **c.** 556   **d.** 72

14. $2 + [3 + (6 - 4)] + 3^2$
    a. 19   **b.** 13   **c.** 90   **d.** 16

15. $\dfrac{2 \cdot 3 + 6}{4 + 2 \cdot 4}$
    a. 12   **b.** 0   **c.** 1   **d.** 4

16. Find the average of 17, 20, 35, and 36.
    a. 24   **b.** 27   **c.** 19   **d.** 108

17. Round 658,125 to the nearest thousand.
    a. 658,000   **b.** 659,000   **c.** 657,000   **d.** 658,100

18. Round each number to the nearest ten and estimate the difference. $79 - 63$
    a. 10   **b.** 20   **c.** 30   **d.** 16

Solve.
19. The product of 4 and 12
    a. 16   **b.** 48   **c.** 8   **d.** 12

20. Find the difference of 87 and 19.
    a. 76   **b.** 72   **c.** 68   **d.** 78

Chapter 1 Practice Test B (cont'd)

21. 18 less 6
    a. 12   **b.** 24   **c.** 108   **d.** 3

22. Find the quotient of 81 and 3.
    a. 84   **b.** 27   **c.** 243   **d.** 78

23. Ruth bought a blouse and a skirt and paid $64. If the blouse cost $38, how much did the skirt cost?
    a. $26   **b.** $102   **c.** $34   **d.** $36

24. A computer cost $694 and a monitor costs $168. How much did they cost together?
    a. $534   **b.** $852   **c.** $862   **d.** $526

25. If one television costs $392, what would 5 televisions cost?
    a. $397   **b.** $387   **c.** $1550   **d.** $1960

26. Three mathematics books cost $225, what would one cost?
    a. $675   **b.** $75   **c.** $64   **d.** $112

27. Find the area of the square.

11 yards

    a. 11 square yards **b.** 22 square yards **c.** 121 square yards **d.** 44 square yards

28. Find the perimeter of the rectangle.

8 inches

3 inches

    a. 22 inches   **b.** 24 square inches   **c.** 121 square inches **d.** 44 inches

Chapter 1 Practice Test B (cont'd)

29. Evaluate $2 + 3(x + 2)^2$ for $x = 3$.
    a. 50   **b.** 125   **c.** 41   **d.** 77

30. Evaluate $3x^2 - 2$ for $x = 4$.
    a. 36   **b.** 16   **c.** 46   **d.** 142

Translate each of the following into a mathematical expression. Use x for "a number."
31. The product of 12 and a number
    a. $\dfrac{12}{x}$ **b.** $12x$ **c.** $12 + x$ **d.** $12 - x$

32. The quotient of 12 and a number
    a. $\dfrac{12}{x}$ **b.** $12x$ **c.** $12 + x$ **d.** $12 - x$

33. Is 2 a solution to $3x - 4 = 1$?
    a. yes   **b.** no

2.1 Introduction to Integers

Learning Objectives
   A. Represent real-life situations with integers.
   B. Graph integers on a number line.
   C. Compare integers.
   D. Find the absolute value of a number.
   E. Find the opposite of a number.
   F. Read bar graphs containing integers.

Key Vocabulary
negative numbers, signed numbers, integers, absolute value, opposite

Objective A Represent real-life situations with integers.

   Example 1 The lowest point in the United States is Badwater Basin in Death Valley at 282 feet below sea level. Represent this position using an integer.

Objective B Graph integers on a number.

   Example 2 Graph −3, 2, 0, 4, −1 on the number line.

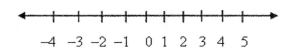

Objective C Compare integers.

   Example 3 Insert < or > between each pair of numbers to make a true statement.

   a) 8 −3   b) −3 0   c) −2 −17

Objective D Find the absolute value of a number.

   Example 4 Simplify.

   a) |7| b) |−8| c) |1|

1) −282, 2) ◄─┼─●─┼─●─┼─┼─●─┼─►   −4 −3 −2 −1 0 1 2 3 4 5   *3a) >, b) <, c) >, 4a) 7, b) 8, c) 1*

Examples 2.1 (cont'd)

Objective E Find opposites.

Example 5 Find the opposite of each number.

a) −12   b) 6   c) 0

Example 6 Simplify.

a) −(2)   b) −(−3)   c) −|−8|

Example 7. Evaluate −|x| if x = −2.

Objective F Read bar graphs containing integers.

The bar graph shows the lowest temperature in the United States for 5 days in January, 2006.

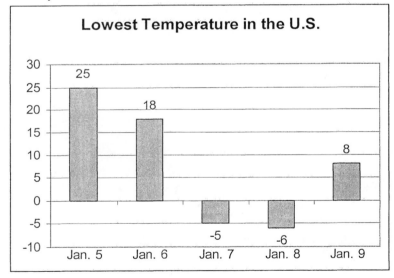

Example 8 Which day had the lowest temperature?

5a) 12, b) −6, c) 0, 6a) −2, b) 3, c) −8, 7) −2, 8) Jan. 8

Practice Set 2.1

Use the choices below to fill in each blank.

absolute value   **integers**
opposites   **signed numbers**

1. The numbers . . . −4, −3, −2, −1, 0, 1, 2, 3, 4, . . . are called _____.

2. The numbers 8 and −8 are _____.

3. _____ are positive numbers, negative numbers, and zero.

4. A number's distance from zero on a number line is its _____.

Objective A   Represent real-life situations with integers.

5. The lowest temperature one night was 8 below zero.   5. _____

6. The Dow Jones gained 2 points one day.   6. _____

Objective B   Graph numbers on a number line.

Graph each integer on the number line.

7. −4, −3, 0, 3   7.
−4 −3 −2 −1 0 1 2 3 4 5

Objective C   Compare integers.
Insert < or > between each pair of integers to make a true statement.

8. 1   −3   8. _____

9. 8   −8   9. _____

10. −4   12   10. _____

Objective D   Find the absolute value of a number.
Simplify.

11. $|-4|$ 11. _____

12. $|1|$ 12. _____

13. $|0|$ 13. _____

Practice Set 2.1 (con't)

Objective E Find the opposite of a number.
Find the opposite of each number.

14. 8   14. _____

15. −6   15. _____

16. 0   16. _____

Simplify.

17. |−18| 17. _____

18. −|3| 18. _____

19. −|−6| 19. _____

Insert <, >, or = between each pairs of numbers to make a true statement.

20. −|−2| 2   20. _____

21. −(−2)   −|−2)   21. _____

Objective F Read bar graphs containing integers.

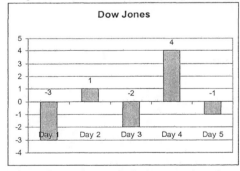

22. What day had the worst loss?   22. _____

Extensions

23. Order the numbers from least to the greatest.   23. _____
     |−4|, −4, −(−3), 1, −(2)

24. True or False: Zero is always less than any   24. _____
     negative integer.

25. Find a value of *a* so that −*a* is a positive integer.   25. _____

2.2 Adding Integers

Learning Objectives
   A. Add integers.
   B. Evaluate an algebraic expression by adding.
   C. Solve problems by adding integers.

Objective A Add integers.

Add using a number line.
   Example 1 $-3 + 4$

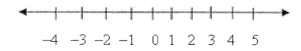

   Example 2 $-2 + (-3)$

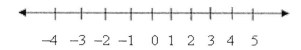

   Example 3 $3 + (-4)$

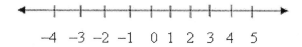

   Example 4 $-4 + 4$

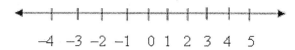

Add.

   Example 5 $-3 + (-6)$

   Example 6 $-5 + (-4)$

   Example 7 $8 + (-4)$

1) 1, 2) –5, 3) –1, 4) 0, 5) –9, 6) –9, 7) 4

Martin-Gay **Prealgebra** edition 5   53

Examples 2.2 (cont'd)

Example 8 $(-5) + 3$

Example 9 $-7 + 10$

Example 10 $15 + (-9)$

Example 11 $(-6) + 0$

Example 12 $20 + (-20)$

Example 13 $-19 + 19$

Example 14 $(-6) + 3 + (-4)$

Example 15 $4 + (-3) + 8 + (-6)$

Objective B Evaluate algebraic expressions by adding.

Example 16 Evaluate $3x + y$ for $x = -3$ and $y = 2$.

Example 17 Evaluate $x + 3y$ for $x = -4$ and $y = -2$.

Objective C Solve problems by adding integers.

Example 18 A science class recorded the outside temperature every hour for an experiment. At 6 a.m. the temperature was 43°, at 7 a.m. the temperature had risen 2°, at 8 a.m. the temperature had fallen 3°, and at 9 a.m. the temperature had fallen 4°. What was the temperature at 9 a.m.?

8) –2, 9) 3, 10) 6, 11) –6, 12) 0, 13) 0, 14) –7, 15) 3, 16) –7, 17) –10, 18) 38°

Practice Set 2.2

1. For any number b, $b + (-b)$ is equal to _____.

Objective A   Add integers.

Add using a number line.

2. $-3 + 4$ 2. _____

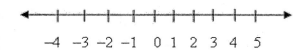

3. $-2 + (-1)$ 3. _____

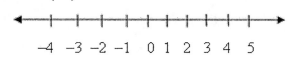

4. $3 + (-2)$ 4. _____

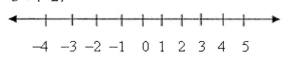

Add.
5. $8 + (-4)$ 5. _____

6. $8 + 6$ 6. _____

7. $12 + (-15)$ 7. _____

8. $-5 + (-12)$ 8. _____

9. $-6 + 4$ 9. _____

10. $-(56) + (-32)$ 10. _____

11. $-5 + 4 + (-3)$ 11. _____

12. $-8 + 17 + (-4) + (-2)$ 12. _____

13. $8 + (-8)$ 13. _____

Practice Set 2.2 (cont'd)

Objective B Evaluate an algebraic expression by adding.

14. Evaluate $3x + y$ for $x = -3$ and $y = 4$.

14. _____

Objective C Solve problems by adding integers.

15. Find the sum of -5 and 8.

15. _____

16. An toy airplane is flying 25 feet high. It descends 8 feet and ascends 10 feet and then descends 5 feet. How high was it after the new maneuvers?

16. _____

A new toy store registered the following earning for its first 5 months in operation.

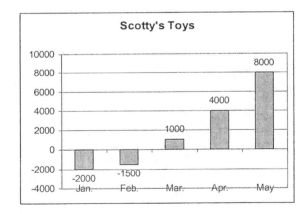

17. What were the total losses for January and February?

17. _____

18. What was the total profit or loss for the five months?

18. _____

Extensions

19. Give two negative numbers whose sum is -8.

19. _____

20. Give one negative number and one positive number whose sum is -8.

20. _____

2.3 Subtracting Integers

Learning Objectives
   A. Subtract integers.
   B. Add and subtract integers.
   C. Evaluate an algebraic expression by subtracting.
   D. Solve problems by subtracting integers.

Objective A Subtract integers.

Subtract.

Example 1 $6 - 3$

Example 2 $-3 - 8$

Example 3 $5 - (-4)$

Example 4 $-8 - (-3)$

Example 5 $-6 - 8$

Example 6 $-8 - (-20)$

Example 7 $3 - 11$

Example 8 Subtract -3 from 8.

1) 3, 2) –11, 3) 9, 4) –5, 5) –14, 6) 12, 7) –8, 8) 11

Name:
Instructor:

Date:
Section:

Examples 2.3 (cont'd)

Objective B Add and subtract integers.

Example 9 Simplify. $9 - 12 - (-3) - 3$

Example 10 Simplify $8 + (-4) - 5 + 4 - (-3)$

Objective C Evaluate and algebraic expression by subtracting.

Example 11 Evaluate $x - y$ for $x = -2$ and $y = -4$.

Example 12 Evaluate $2a - b$ for $x = 3$ and $y = -3$.

Objective D Solve problems by subtracting integers.

Example 13 The lowest point in Death Valley National Park is Badwater Basin at 282 feet below sea level. The highest point in Death Valley National Park is Telescope Peak at 11,004 feet above sea level. How much higher is Telescope Peak than Badwater Basin?

9) –3, 10) 6, 11) 2, 12) 9, 13) 11,286 feet

Practice Set 2.3

Objective A Subtract integers.

Find each difference.

1. $-5 - (-2)$

1. _____

2. $-4 - (-4)$

2. _____

3. $8 - 4$

3. _____

4. $-8 - 3$

4. _____

5. $-6 - (-12)$

5. _____

6. $8 - (-2)$

6. _____

7. $9 - 12$

7. _____

8. $7 - 8$

8. _____

9. $-5 - (-2)$

9. _____

10. $-148 - (-200)$

10. _____

11. Subtract 6 from -3.

11. _____

12. Subtract -8 from -10.

12. _____

Objective B Add and subtract integers.

13. $5 - 3 - 4$

13. _____

14. $-12 - (-2) + 6 - 8$

14. _____

15. $-5 - (-3) + 5$

15. _____

Practice Set 2.3 (cont'd)

16. $-(-3) -4 - 8 - 10$ 16. _____

17. $8 + 6 - 4 - 12$ 17. _____

18. $-6 - 8 + 10 + 20$ 18. _____

Objective C Evaluate an algebraic expression by subtracting.

19. Evaluate $x - y$ for $x = 3$ and $y = -2$.   19. _____

20. Evaluate $4x - y$ for $x = -2$ and $y = -1$.   20. _____

Objective D Solve problems by subtracting integers.

21. Julia made 3 points below average on her first test and   21. _____
     5 points above average on her second test. How much
     did she improve on her second test?

22. The hottest temperature ever recorded in one small town   22. _____
     in Texas was 112°. The lowest temperature was 8° below
     zero. What was the difference between the highest and
     the lowest temperature?

Extensions

23. Give two numbers whose difference is -8.   23. _____

24. Give one negative and one positive number whose   24. _____ .
     difference is -8.

2.4 Multiplying and Dividing Integers

Learning Objectives
    A. Multiply integers.
    B. Divide integers.
    C. Evaluate an algebraic expression by multiplying and dividing.
    D. Solve problems by multiplying or dividing integers.

Objective A Multiply integers.

Multiply.
Example 1 $-6 \cdot 4$

Example 2 $-3(-5)$

Example 3 $0 \cdot (-8)$

Example 4 $12(-3)$

Example 5 $3(-4)(-2)$

Example 6 $(-4)(-3)(-5)$

Example 7 $(-3)(-4)(-1)(-2)$

Example 8 Evaluate $(-3)^2$

1) −24, 2) 15, 3) 0, 4) −36, 5) 24, 6) −60, 7) 24, 8) 9

Examples 2.4 (cont'd)

Example 9 Evaluate -3^2

Objective B Divide integers.

Divide.

Example 10 $\dfrac{-8}{4}$

Example 11 $-18 \div (-6)$

Example 12 $\dfrac{36}{-9}$

Divide, if possible.

Example 13 $\dfrac{0}{-6}$

Example 14 $\dfrac{-5}{0}$

Objective C Evaluate an algebraic expression by multiplying and dividing.

Example 15 Evaluate xy for $x = -3$ and $y = 2$.

Example 16 Evaluate $\dfrac{x}{y}$ for $x = -8$ and $y = -2$.

Objective D Solve problems by multiplying or dividing integers.

Example 17 Mary was doing an experiment in chemistry. She had to cool a mixture. She measured the temperature every ten minutes for 6 times. She found the mixture had cooled 2 degrees each time. What was the total loss in the temperature?

9) –9, 10) –2, 11) 3, 12) –4, 13) 0, 14) undefined, 15) –6, 16) 4, 17) –12°

Practice Set 2.4

Use the choices below to fill in each blank.

negative   **positive**   **0**   **undefined**

1. The product of a positive and a negative number is _____.

2. The product of two negative numbers is _____.

3. The quotient of a positive and a negative number is _____.

4. The quotient of two negative numbers is _____.

5. The quotient of 0 and a negative number is _____.

6. The quotient of a negative number and 0 is _____.

Objective A Multiply integers.

Multiply.

7. $-3(-6)$ 10. _____

8. $8(-5)$ 12. _____

9. $-3(0)$ 13. _____

10. $3(-5)(-4)$ 10. _____

11. $-5(-3)(-2)(-2)$ 11. _____

12. $-3(-1)(4)(0)$ 12. _____

Evaluate.

13. -4^2 13. _____

14. $(-4)^2$ 14. _____

Practice Set 2.4 (cont'd)

Objective B Divide integers.

15. $-12 \div (-3)$

15. _____

16. $80 \div (-5)$

16. _____

17. $\dfrac{-42}{8}$

17. _____

18. $\dfrac{-12}{0}$

18. _____

Objective C Evaluate an algebraic expression by multiplying and dividing.

19. Evaluate xy for $x = -2$ and $y = 4$.

19. _____

20. Evaluate $\dfrac{x}{y}$ for $x = -12$ and $y = 12$.

20. _____

Objective D Solve problems by multiplying and dividing.

21. Find the product of -8 and -6.

21. _____

22. Find the quotient of -32 and -16.

22. _____

23. A football team lost 4 yards on each of 3 consecutive plays. Find the total yards lost.

23. _____

Extensions

24. What is the sign of the product of 13 negative numbers?

24. _____

25. What is the sign of the product of 28 negative numbers?

25. _____

26. Evaluate $(-1)^{129}$

26. _____

27. Evaluate -1^{130}

27. _____

2.5 Order of Operations

Objective A Simplify expressions by using the order of operations.

Find the value of each expression.

Example 1 $(-5)^2$

Example 2 -2^2

Example 3 $3 \cdot (-2)^3$

Simplify.

Example 4 $\dfrac{2(-6)}{-4}$

Example 5 $\dfrac{18-12}{-3-3}$

Example 6 $20-15+(-4)^3$

Example 7 $-2^4+(-2)^4-1^5$

Example 8 $3-5(4-6)-(-4)$

1) 25, 2) –4, 3) –24, 4) 3, 5) –1, 6) –59, 7) –1, 8) 17

Examples 2.5 (cont'd)

Example 9 $(-6)(-5) - (-3) + (-2)^2$

Example 10 $-4[-6 + 2(-5 + 8)] - 12$

Objective B Evaluate an algebraic expression.

Example 11 Evaluate x^2 and $-x^2$ for $x = -10$.

Example 12 Evaluate $3x^2$ for $x = -4$.

Example 13 Evaluate $3x - 2y + z$ for $x = 3$, $y = -2$, and $z = 12$.

Evaluate 14 Evaluate $12 - x^2 + 2x$ for $x = -3$.

Objective C Find the average of a list of numbers.

Example 15 For one week, the Dow Jones average fell 3 points, rose 2 points, fell 5 points, rose 3 points, and fell 7 points. What was the average loss or gain for the week?

9) 37, 10) –12, 11) 100, –100, 12) 48, 13) 25, 14) –3, 15) –2°

Practice Set 2.5

Use the choices below to fill in each blank.

addition　　　**subtraction**　　　**multiplication**　　　**division**

1. To simplify $3 + 8 \cdot (-5)$, which operation should be performed first? _____

2. To simplify $12 \div 3 \cdot 2$, which operation should be performed first? _____

3. To simplify $\dfrac{5+1}{-3}$, which operation should be performed first? _____

4. To simplify $3 - (-5) + 4$, which operation should be performed first? _____

Objective A Simplify expressions.

Simplify.

5. $7 + 18 \div (-3)$　　　　　　　　　　5. _____

6. $3 + 2(8 - 12)$　　　　　　　　　　6. _____

7. $\dfrac{20 - (-4)}{3 - (-3)}$　　　　　　　　　　7. _____

8. $3 \cdot (-2) - (4 - 8) + 5 \cdot (2)^2$　　　　　8. _____

9. $|-5 - 3| + 2^2$　　　　　　　　　　9. _____

10. $-(-4)^3$　　　　　　　　　　10. _____

11. $(7 - 4) \div (9 - 12)$　　　　　　　11. _____

Practice Set 2.5 (cont'd)

12. $(-13 - 3) \div (-4) \cdot 2$

12. _____

13. $-2\left(3 - \sqrt{49}\right) - 4(1 - 5)^2$

13. _____

Objective B Evaluate an algebraic expression.

Evaluate.

14. $x - y + z$ for $x = -1$, $y = -3$, and $z = -4$.

14. _____

15. $x^3 - x^2$ for $x = -3$.

15. _____

16. $3x^2$ for $x = -5$.

16. _____

Objective C Find the average of a list of numbers.

17. Find the average: $-3, -8, 2, -5, 8, -6$

17. _____

18. Find the average: $-25, -20, 20, -15, -35$

18. _____

19. Mr. Rogers scored 6 below par one day at golf and 2 above par the next day. Find the difference between the two scores.

19. _____

Extensions
20. Are parentheses necessary in the expression $(3 + 2)4$? Explain your answer.

20. _____

2.6 Solving Equations: the Addition and Multiplication Properties

Learning Objectives
    A. Identify solutions of equations.
    B. Use the addition property of equality to solve equations.
    C. Use the multiplication property of equality to solve equations.

Objective A Divide whole numbers.

    Example 1 Determine if -2 is a solution to the equation $3x + 4 = 2$.

Objective B Use the addition property of equality to solve equations.

    Example 2 Solve: $x + 4 = -3$ for x.

    Example 3 Solve: $-5 = x - 3$.

    Example 4 Solve: $10x = 9x - 4$

1) no, 2) −7, 3) −2, 4) −4

Examples 2.6 (cont'd)

Objective C Use the multiplication property of equality to solve equations.

    Example 4 Solve: $3x = -12$

    Example 5 Solve: $-2x = -10$

    Example 6 Solve: $-15 = -5x$

    Example 7 Solve: $-15x = 45$

    Example 8 $\dfrac{x}{-2} = -4$

4) –4, 5) 5, 6) 3, 7) –3, 8) 8

Practice Set 2.6

Use the choices below to fill in each blank.

equation	multiplication	addition
expression	solution	equivalent

1. $3x + 6$ is a(n) _____ and $3x = 6$ is a(n) _____.

2. If the same number is added to or subtracted from both sides of an equation, the _____ property of equality has been used.

3. If the same nonzero number has been multiplied or divided on both sides of the equation, the _____ property of equality has been used.

4. Equations that have the same solution are said to be _____.

5. A(n) _____ can be simplified or evaluated while a(n) _____ can be solved.

6. A number that can be substituted for a variable in an equation and makes a true statement is a _____.

Objective A Identify solutions of equations.

7. Is 5 a solution of $2x + 3 = 15$?   7. _____

8. Is -3 a solution of $x - 3 = -6$?   8. _____

Objective B Use the addition property of equality to solve equations.

9. $a + 4 = -10$ 9. _____

10. $b - 6 = 8$ 10. _____

Practice Set 2.6 (cont'd)

11. $3 = x + 8$

11. _____

12. $12x = 11x - 4$

12. _____

Objective C Use the multiplication property of equality to solve.

13. $4x = 28$

13. _____

14. $-2y = -8$

14. _____

15. $\dfrac{x}{-4} = -2$

15. _____

16. $\dfrac{x}{3} = -9$

16. _____

Solve.

17. $-25 = -17 + x$

17. _____

18. $\dfrac{x}{-4} = 0$

18. _____

19. $\dfrac{x}{-10} = -20$

19. _____

20. $5x = 4x - 8$

20. _____

Extensions

21. Write an equation that can be solved using the addition property of equality. Solve your equation.

21. _____

Chapter 2 Vocabulary Reference Sheet

Term	Definition	Example
Section 2.1		
Positive number	Any number to the right of zero on a number line.	6, 10, + 3 are positive numbers
Negative number	Any number to the left of zero on a number line.	−6, −12 are negative numbers
Signed numbers	Positive numbers, negative numbers and zero	−6, 0, 3 are signed numbers
Inequality signs	$<, >, \geq, \leq$	8 > 3 is read "8 is greater than 3" 4 < 7 is read "4 is greater than 7"
Absolute value	The number's distance from zero on the number line.	|3| is 3, |−3| is 3
Opposite	2 numbers that are the same distance from zero on the number, but in opposite directions.	8 and −8 are opposites.
Integers	The numbers labeled on a number line.	. . . −3, −2, −1, 0, 1, 2, 3, . .
Section 2.2		
Equivalent equations	Equations that have the same solution.	$x + 2 = 4$ and $x = 2$ are equivalent equations.

NOTES:

Chapter 2 Practice Test A

Simplify each expression.

1. $-6 + 5$ **1.** _____

2. $7 - 14$ **2.** _____

3. $2(-5)$ **3.** _____

4. $(-8) \div (-2)$ **4.** _____

5. $18 \div (-3)$ **5.** _____

6. $-9 - (-8)$ **6.** _____

7. $5(-12)$ **7.** _____

8. $\dfrac{-100}{20}$ **8.** _____

9. $|-17| - (-12)$ **9.** _____

10. $13 - |-2|$ **10.** _____

11. $|3| \cdot |-2|$ **11.** _____

12. $\dfrac{|-20|}{-20}$ **12.** _____

Chapter 2 Practice Test A (cont'd)

13. $-7 + 5 \div (-5)$

13. _____

14. $-8 + (-2) - 3 + 6$

14. _____

15. $(-2)^3 + 8 \div (-2)$

15. _____

16. $(6-2)^2 \cdot (5-9)^2$

16. _____

17. $-(-2)^2 \div 4(-2)$

17. _____

18. $2 - 3(4-2)^2$

18. _____

19. $\dfrac{8}{4} - \dfrac{2^4}{4}$

19. _____

Chapter 2 Practice Test A (cont'd)

20. $\dfrac{-(5)(-2)+5}{-9+6}$

20. _____

21. Evaluate $2x - 3y$ for $x = -3$ and $y = 2$.

21. _____

22. Evaluate $4 - x^2$ for $x = -2$.

22. _____

23. Evaluate $\dfrac{8x}{4y}$ for $x = -1$ and $y = -2$.

23. _____

24. Juanita has $425 in her checking account. She writes a check for $45, withdraws $60 from an ATM and deposits $125. Represent the new balance as an integer.

24. _____

Chapter 2 Practice Test A (cont'd)

25. Joyce dives 37 feet down from sea level, raises 12 feet   25. _____
    and then dives down 42 feet. Represent her final depth as
    an integer.

26. The highest point in one country is 43,953 feet above   26. _____
    sea level. The lowest point is 20 feet below sea level.
    Represent the difference in elevation of these two places
    by an integer.

27. Find the average of −8, −12, and 26.   27. _____

Solve.
28. $-3x = 15$ 28. _____

29. $\dfrac{x}{-4} = -8$ 29. _____

30. $x + 8 = 6$ 30. _____

31. $5x = 4x - 2$ 31. _____

Chapter 1 Practice Test B

Simplify.

1. $-9 + 4$
    a. 5   **b.** -5 **c.** 13   **d.** -13

2. $10 - 15$
    a. 25   **b.** -25 **c.** -5 **d.** 5

3. $-4(-6)$
    a. 24   **b.** -24 **c.** -10 **d.** -2

4. $(-49) \div 7$
    a. 7   **b.** -7 **c.** -56 **d.** -42

5. $(-18) + (-6)$
    a. 12   **b.** -12 **c.** 24   **d.** -24

6. $-12 - (-8)$
    a. -20 **b.** 20   **c.** -4 **d.** 4

7. $(-3)(8)$
    a. -24 **b.** 24   **c.** 5   **d.** -5

8. $\dfrac{-20}{5}$
    a. 4   **b.** -4 **c.** -15 **d.** -25

9. $|-12| + |-3|$
    a. 15   **b.** -15 **c.** -9 **d.** 9

10. $17 - (-8)$
    a. 25   **b.** 9   **c.** -9 **d.** 11

Chapter 2 Practice Test B (cont'd)

11. $3 \cdot |-2|$

    a. -1 **b.** 1 **c.** -6 **d.** 6

12. $\dfrac{|-30|}{-|-6|}$

    a. 5 **b.** -5 **c.** 24 **d.** -36

13. $\dfrac{-12}{0}$

    a. 0 **b.** -12 **c.** 12 **d.** undefined

14. $\dfrac{0}{-7}$

    a. 0 **b.** -7 **c.** 7 **d.** undefined

15. $-3 + (-2) + 6 - 5$

    a. 0 **b.** -6 **c.** 2 **d.** -4

16. $-2^2 + (-2)^2$

    a. -8 **b.** 0 **c.** 8 **d.** 16

17. $5 - 2(3 - 6)$

    a. -9 **b.** 9 **c.** 11 **d.** -1

18. $(4 - 6)^2 \cdot (3 - 7)^2$

    a. -64 **b.** 64 **c.** 20 **d.** 12

19. $\dfrac{-5^2}{-5}$

    a. -1 **b.** -5 **c.** 5 **d.** 1

Chapter 2 Practice Test B (cont'd)

20. $\dfrac{(-8)(-2)-4}{-1(8-5)}$

    a. −4   **b.** 4   **c.** $\dfrac{20}{3}$ **d.** $-\dfrac{20}{3}$

21. $\dfrac{-|8-10|}{3^2-2\cdot4}$

    a. −18   **b.** 18   **c.** 2   **d.** −2

22. $3(-2)-\left[4-(-2-3)^2\right]$

    a. 15   **b.** 3   **c.** −15   **d.** −3

23. Evaluate $2x-3y$ for $x=2$ and $y=-3$.

    a. 5   **b.** −5   **c.** 13   **d.** −13

24. Evaluate $6-x^2$ for $x=-3$.

    a. −15   **b.** 3   **c.** 15   **d.** −3

25. Evaluate $\dfrac{2a}{3b}$ for $a=-6$ and $b=-2$.

    a. −2   **b.** 2   **c.** −3   **d.** 4

26. For 4 weeks in a row, Tanner wrote a check for $27 for gasoline. Represent the change to his account by an integer.

    a. −108   **b.** 108   **c.** −23   **d.** −54

27. From sea level, Frank dove 36 feet, rose 20 feet and dove 8 feet. Represent his final depth as an integer.

    a. 64   **b.** −8   **c.** −24   **d.** −64

Chapter 2 Practice Test B (cont'd)

28. The highest point in Louisiana is Driskill Mountain at 535 feet. The lowest point is New Orleans at 6 feet below sea level. Represent the difference in elevations as an integer. (*Source:* U.S. Geographical Survey)
    a. 541   **b.** 539   **c.** 3218   **d.** 72

29. Find the average of −16, −8, 4, and 12.
    a. −10   **b.** 10   **c.** 2   **d.** −2

Solve.

30. $3x = -21$
    a. −24   **b.** −18   **c.** −7   **d.** 7

31. $x - 6 = -10$
    a. 4   **b.** −4   **c.** −16   **d.** 16

32. $\dfrac{x}{-8} = -8$
    a. 64   **b.** 1   **c.** −64   **d.** −1

33. $5x = 6x - 2$
    a. $\dfrac{2}{11}$ **b.** $-\dfrac{2}{11}$ **c.** −2   **d.** 2

Name: **Date:**

Instructor: **Section:**

3.1 Simplifying Algebraic Expressions

Learning Objectives
 A. Use properties of numbers to combine like terms.
 B. Use properties of numbers to multiply expressions.
 C. Simplify expressions by multiplying and then combining like terms.
 D. Find the perimeter and area of figures.

Key Vocabulary
numerical coefficient, like terms, unlike terms

Objective A Use properties of numbers to combine like terms.

  Example 1 Simplify each expression by combining like terms.

  (a) $5x + 8x$ (b) $3y - 5y$ (c) $8x^2 - 7x^2 + 4x$

  Example 2 Simplify: $3x - 4 + 4x - 8$

  Example 3 Simplify: $4x - 5x - 8$

  Example 4 Simplify: $7x - 8 + 3x - 12$

  Example 5 Simplify: $5x - 3y + 4 - 4x + 2y - 5$

Objective B Use properties of numbers to multiply expressions.

  Multiply.

  Example 6 $-2(3x)$

  Example 7 $3(6x)$

  Example 8 $3(x + 4)$

  Example 9 $-3(4b - 5)$

  Example 10 $6(x - 2)$

1a) 13x, b) –2y, c) $x^2 + 4x$, 2) 7x–12, 3) –x – 8, 4) 10x – 20, 5) x – y – 1, 6) –6x, 7) 18x,
8) 3x + 12, 9) –12b + 15, 10) 6x – 12

Martin-Gay **Prealgebra** edition 5

Examples 3.1 (cont'd)

Objective C Simplify expressions by multiplying and then combining like terms.

Simplify.
Example 11 $2(3 - 2x) - 5$

Example 12 $-3(x + 2) + 4(3x - 5)$

Example 13 $3 - (x + 2)$

Objective D Find the perimeter and area of figures.

Example 14 Find the perimeter of the triangle.

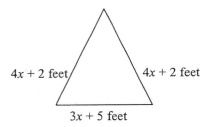

$4x + 2$ feet $4x + 2$ feet

$3x + 5$ feet

Example 15 Find the area of the rectangle.

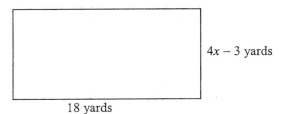

$4x - 3$ yards

18 yards

11) −4x + 1, 12) 9x − 26, 13) −x + 1, 14) 11x + 9 feet, 15): 72x − 54 yards

Practice Set 3.1

Use the choices below to fill in each blank. Some terms will be used more than once.

combine like terms   **like**   **variable**   **unlike**

1. We _____ to simplify $3x + 4x$.

2. In the term $-3x$, -3 is called the _____ and x is called the
   _____.

3. The term a has a _____ of 1.

4. The terms $3x$ and $4y$ are _____ terms and $5x$ and $-2x$ are
   _____ terms.

Objective A Use properties of numbers to combine like terms.

Simplify each expression by combining like terms.
5. $4x + 6x$ 5. _____

6. $3a - a + 4a$ 6. _____

7. $3y + 4y - 8y + 3$ 7. _____

Objective B Use properties of numbers to multiply.

Multiply.
8. $3(4x)$ 8. _____

9. $-3(2c)$ 9. _____

10. $-7(-8b)$ 10. _____

11. $3(x + 4)$ 11. _____

12. $-5(4x - 3)$ 12. _____

Objective C Simplify expressions by multiplying and then combining like terms.

Simplify each expression.
13. $2(x - 4) + 6$ 13. _____

14. $2(3x - 5) + 2(x - 4)$ 14. _____

Practice Set 3.1 (cont'd)

15. $3x - 5 + 5x - 8$

15. _____

16. $-3(x - 3) + 6(5x - 3)$

16. _____

17. $-2(3b - 4) - 6b$

17. _____

18. $-(4xy + 21) + 2(3xy - 7)$

18. _____

Objective D Find the perimeter and area of figures.

19. Find the perimeter of the figure.

19. _____

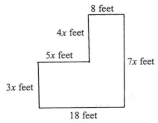

8 feet

4x feet

5x feet

7x feet

3x feet

18 feet

20. Find the area of the rectangle.

20. _____

2x inches

25 inches

21. Find the area of a rectangular floor that is 20 feet by 35 feet.

21. _____

22. Find the perimeter of a rectangular floor that is 20 feet by 35 feet.

22. _____

Extensions

23. Write an expression that represents the area of the composite figure. Then simplify to find the total area.

23. _____

8 feet

4x feet

3x feet

18 feet

3.2 Solving Equations: Review of the Addition and Multiplication Properties

Learning Objectives
   A. Use the addition property or the multiplication property to solve equations.
   B. Use both properties to solve equations.
   C. Translate word phrases to mathematical expressions.

Vocabulary
simplify

Objective A Use the addition property or the multiplication to solve equations.

Example 1 Solve: $x - 4 = 8 - 12$

Example 2 Solve: $7y - 4y = 9$

Example 3 Solve: $\dfrac{x}{2} = 4 - 7$

Example 4 Solve: $5x - 3x + 2 = 17 - 9$

Example 5 Solve: $7x - 8x = 8 - 3$

Example 6 Solve: $2(3x - 10) = 5x$

1) 0, 2) 3, 3) –6, 4) 3, 5) –5, 6) 20

Examples 3.2 (cont'd)

Objective B Use both properties to solve equations.

Example 7 Solve: $3x + 2 = 23$

Example 8 $5 - 6 = -2(x + 2) - 3$

Objective C Translate word phrases into mathematical expressions.

Example 9 Write each phrase as an algebraic expression. Use x to represent "a number."

(a) a number increased by 6

(b) the product of a number and -4

(c) the quotient of 2 and a number

(d) a number subtracted from 6

(e) 4 less than a number

Example 10 Write each phrase as an algebraic expression. Let x be the unknown number.

(a) three times a number, increased by 7

(b) four times the sum of a number and -2

(c) the sum of 8 and a number, divided by 3

7) 7, 8) –3, 9a) x + 6, b) –4x, c) $\dfrac{2}{x}$, d) 6 – x, e) x – 4, 10a) 3x+7, b) 4(x-2), c) $\dfrac{x+8}{3}$,

Practice Set 3.2

1. If you write $3x + 4x$ as $7x$, what do you call the operation?_____

Objective A Use the addition property and multiplication property to solve equations.

Solve.

2. $-4x = 17 - 5$

2. _____

3. $2 - 7 = \dfrac{a}{-2}$

3. _____

4. $-8 - 6 = 7x - 2 - 6x$

4. _____

5. $-5 + 2 = 2x - 8 - x$

5. _____

6. $2(3x - 1) = 4x - 6$

6. _____

Objective B Use both properties to solve equations.

Solve.

7. $5x + 15 = 0$

7. _____

8. $2(5y - 6) = 6y$

8. _____

9. $3(x - 7) = 10 - 19$

9. _____

Practice Set 3.2 (cont'd)

10. $7 - 49 = 2(3x + 4) - 2$ **10.** _____

11. $12 - 8 = \dfrac{x}{-5}$ **11.** _____

12. $82 - 62 = 10x - 15x$ **12.** _____

13. $7x - 4x = 20 - 4(5)$ **13.** _____

14. $\dfrac{x}{3} = -7 - (-4)$ **14.** _____

Objective C Translate word phrases to mathematical expressions.
Write each phrase as a mathematical expression. Use x to represent "a number."

15. The sum of 9 and a number   **15.** _____

16. Seventeen subtracted from a number   **16.** _____

17. The product of a number and -3 **17.** _____

18. Negative 8 divided by a number   **18.** _____

19. The product of 8 and a number, added to 25   **19.** _____

Extensions

20. Is the equation $-x = -3$ solved? If not, solve it. **20.** _____

Martin-Gay **Prealgebra** edition 5   90

3.3 Solving Linear Equations in One Variable

> **Learning Objectives**
>   A. Solve linear equations using the addition and multiplication properties.
>   B. Solve linear equations containing parentheses.
>   C. Write numerical sentences as equations.

Objective A Solve linear equations using the addition and multiplication properties.

Example 1 Solve: $4a - 2 = 2a + 6$

Example 2 Solve: $-14 - 3x + 2 = 2x - x$

Objective B Solve linear equations containing parentheses.

Example 3 Solve: $5(x - 8) = 7(x - 2)$

1) 4, 2) –3, 3) –13

Examples 3.3 (cont'd)

Example 4 Solve: $7(3x-4)+70=0$

Objective C Write numerical sentences as equations.

Example 5 Translate each sentence into an equation.

(a) The difference of 12 and 5 is 7.

(b) Twice the sum of 8 and −3 is 10.

(c) The quotient of −72 and −9 is 8.

4) −2, 5a) 12 − 5 = 7, b) 2(8− 3) = 10, c) $\dfrac{-72}{-9} = 8$

Practice Set 3.3

Objective A Solve linear equations using the addition and multiplication properties.
Solve.

1. $7x - 8 = 5x + 12$ 1. _____

2. $18 - 9x = 8 - 4x$ 2. _____

3. $8 - 7x = -2x + 8$ 3. _____

4. $20x - 50 = 10x + 100$ 4. _____

5. $8x - 11 - 6x = 2x - 8 + 3x + 6$ 5. _____

Practice Set 3.3 (cont'd)

Objective B Solve linear equations containing parentheses.

6. $3(x+4) = 20-11$

6. _____

7. $4(x-7)+16 = 0$

7. _____

8. $2(x-4) = x+17$

8. _____

9. $4-(y-8) = 20$

9. _____

10. $4(7x-3) = 12(x-5)$

10. _____

Practice Set 3.3 (cont'd)

11. $-x = 8$ **11.** _____

12. $7 - b = 20$ **12.** _____

13. $5y + 8 = -42$ **13.** _____

14. $2(7a - 3) = 10(2a - 5) - 4$ **14.** _____

15. $4(x + 2) = 7x + 8$ **15.** _____

16. $17 + 7(w + 2) = 3w - 5$ **16.** _____

Practice Set 3.3 (cont'd)

Objective C   Write numerical sentences as equations.

17. The sum of -8 and -12 is -20.

17. _____

18. The product of 12 and -3 is -36.

18. _____

19. Four times the difference of 8 and 5 is 12.

19. _____

20. Eight subtracted from 11 is 3.

20. _____

Extensions

21. A classmate shows you his steps for solving the given equation. His solution does not check, but he is unable to find his error. Find the error and correct it to find the correct solution.

21. _____

$$-(2x - 8) = -4x + 12$$
$$-2x - 8 = -4x + 12$$
$$-2x + 4x - 8 = -4x + 4x + 12$$
$$2x - 8 = 12$$
$$2x - 8 + 8 = 12 + 8$$
$$2x = 20$$
$$\frac{2x}{2} = \frac{20}{2}$$
$$x = 10$$

3.4 Solving Equations: the Addition and Multiplication Properties

Learning Objectives
   A. Write sentences as equations.
   B. Use problem-solving steps to solve problems.

Objective A Write sentences as equations.

   Example 1 Write each sentence as an equation. Use x to represent "a number."

   (a) Fifteen increased by a number is 25.

   (b) Four times a number is -24.

   (c) A number minus 8 and 4.

   (d) Four times the sum of a number and -3 is 12.

   (e) The quotient of three times a number and 4 is 15.

Objective B Use problem-solving steps to solve problems.

   Example 2 Twice a number plus 5 is 19. Find the unknown number.

1a) $15 + x = 25$, b) $4x = -24$, c) $x - 8 = 4$, d) $4(x - 3) = 12$, e) $\dfrac{3x}{4} = 15$, 2) 7

Examples 3.4 (cont'd)

Example 3 John drives from his home to school and then to work. His total drive one way is 100 miles. The drive to work was 20 miles more than the drive to school. How long was each drive?

Example 4   Jan spent $85 at a music store. She spent 4 times as much on CD's than she did on a music video. How much did she spend on CD's?

3) 40 miles to school, 60 miles to work 4) $68 on CD's

Practice Set 3.4

Objective A Write sentences as equations.

Write each sentence as an equation. Use x to represent "a number."

1. A number added to -4 is 8.   1. _____

2. The quotient of 7 and a number is -4.   2. _____

3. A number subtracted from 8 is -25.   3. _____

4. Four times a number gives 220.   4. _____

5. Twice the number added to -8 is 24.   5. _____

6. Four times the sum of a number and -3 is 36.   6. _____

Objective B Use problem-solving steps to solve problems.
Translate each problem to an equation. Then solve the equation.

7. Four times a number, added to 7 is 19. Find the number.   7. _____

8. The sum of -5, -3 and a number is -12. Find the number.   8. _____

9. The difference of a number and 7 is equal to 17. Find the number.   9. _____

10. Eight times the difference of some number and 5 is 32. Find the number.   10. _____

11. Eight subtracted from three times a number is 6 more than the number. Find the number.   11. _____

12. Seven times some number divided by 5 is 14. Find the number.   12. _____

Practice Set 3.4 (cont'd)

13. The product of five and a number is the same as the number increased by 44. Find the number.

13. _____

14. Professor has a total of 48 college algebra students in two classes. The morning class has 4 more students than the afternoon class. How many students are in the morning class?

14. _____

15. Frank drives 5 mph faster than Ruth. The total of their speeds is 125 mph. How fast did each drive?

15. _____

16. The average price of a CD is 7 dollars less than the average price of a DVD. The sum of the two average prices is $33. What is the average price of each?

16. _____

17. The city manager is doing a study of traffic at two different intersections in his town. Corner A has 3 times the traffic that Corner B has. The total traffic at both intersections is 232 cars. How many cars were at Corner A?

17. _____

18. Barbara is buying coats and sweaters for a charity to hand out at Christmas. She bought five times as many coats as sweaters. If she bought a total of 72 items, how many coats and how many sweaters did she buy?

18. _____

Extensions

19. The profit P a retailer makes on an item can be computed using the equation $P = S - C$, where S is the selling price and C is the cost the retailer paid. If a retaierl made a profit of $16 one item and the selling price was $34, what was the retailer's cost?

19. _____

Chapter 3 Vocabulary Reference Sheet

Term	Definition	Example
	Section 3.1	
Variable	A letter used to replace a number.	$2x + 3$, x is a variable
Terms	The addends of an expression.	In $4x + 5y$, $4x$ and $5y$ are terms.
Numerical coefficients	The number factor of a variable term.	In $7x$, 7 is the numerical coefficient.
Like terms	Terms that are exactly the same except they have different numerical coefficients.	$3x$ and $-5x$ are like terms. $4x$ and $7y$ are not like terms.
Constant	A term that is a number only.	In the expression $5x + 8$, 8 is the constant.
Simplified	An algebraic expression is simplified if all like terms have been combined.	$3x - 6 + 4x + 4$ simplifies to $7x - 2$.

NOTES:

Name: Date:
Instructor: Section:

Chapter 3 Practice Test A

Simplify each expression.

1. $3x + 8 - 7x + 8$ 1. _____

2. $-4(2x - 3)$ 2. _____

3. $3x - 4 - (x + 2)$ 3. _____

4. Write an expression that represents the perimeter   4. _____
of the triangle. Simplify the expression.

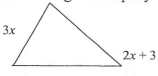

5. Write an expression that represents the area of   5. _____
the rectangle.

$4x - 3$

5

Solve each equation.

6. $7 = 8a - 15a$ 6. _____

7. $\dfrac{x}{3} = 7 - 4$ 7. _____

Chapter 3 Practice Test A (cont'd)

8. $4x + 8 - 2x + 6 = 12$

8. _____

9. $-3x + 7 = -14$

9. _____

10. $3(x + 4) = 6$

10. _____

11. $-3(x + 5) - 7 = 5$

11. _____

12. $7x - 8 = 6x + 4$

12. _____

13. $3 + 2(2x - 8) = 15$

13. _____

14. $5(x - 3) + 5 = 3(x - 7) + 5$

14. _____

Chapter 3 Practice Test A (cont'd)

Translate the following phrases into mathematical expressions. Use x to represent "a number."

15. The sum of −6 and a number

15. _____

16. Four times a number subtracted from 7.

16. _____

Translate each sentence into an equation. Use x to represent "a number."

17. The sum of three times a number and 9 is −7.

17. _____

18. Six added to 4 times a number equals 25.

18. _____

Solve.

19. Eight subtracted from twice a number is 8. Find the number.

19. _____

20. In a college algebra class, there are 5 more females than males. If there are 37 in the class, how many males and how many females are there?

20. _____

Chapter 3 Practice Test B

Simplify each expression.

1. $7x - 8 - 3x + 4$

    a. $4x - 12$ **b.** $4x - 4$ **c.** $4x + 4$ **d.** $10x - 4$

2. $5(3x - 2)$

    a. $15x - 2$ **b.** $8x - 2$ **c.** $15x - 3$ **d.** $15x - 10$

3. $4x - 8 - (2x - 5)$

    a. $2x - 3$ **b.** $2x - 13$ **c.** $6x - 13$ **d.** $6x - 3$

4. Write an expression that represents the perimeter of the triangle. Simplify the expression.

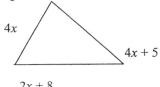

    a. $6x + 13$ **b.** $6x + 8$ **c.** $10x + 13$ **d.** $10x + 5$

5. Write an expression that represents the area of the rectangle. Simplify the expression.

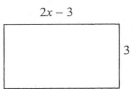

    a. $6x - 9$ **b.** $2x - 6$ **c.** $6x - 3$ **d.** $2x$

Solve each equation.

6. $18 = 5x + 4x$

    a. -2 **b.** 2 **c.** -4 **d.** 4

7. $\dfrac{x}{-3} = 4 - 6$

    a. 6 **b.** -6 **c.** -2 **d.** 4

Chapter 3 Practice Test B (cont'd)

8. $4x + 5 - 3x - 4 = 21$

    a. −3   **b.** 3   **c.** 11   **d.** 20

9. $-5x + 4 = -31$

    a. 7   **b.** −7   **c.** 31   **d.** 5

10. $6(x - 3) = 18$

    a. 0   **b.** 3   **c.** 6   **d.** −3

11. $-2(x - 2) - 5 = 11$

    a. −3   **b.** −6   **c.** 9   **d.** −9

12. $11x - 7 = 10x + 4$

    a. 3   **b.** −11   **c.** 11   **d.** 7

13. $4 - 3(x - 7) = 13$

    a. 4   **b.** −4   **c.** 16   **d.** −7

14. $8 - 3(x + 2) = 7x - 8$

    a. −1   **b.** 2   **c.** 10   **d.** 1

Translate the following phrases into mathematical expressions. Use x to represent "a number."

15. The sum of 8 and a number

    a. $8x$ **b.** $x + 8$ **c.** $\dfrac{x}{8}$ **d.** $x - 8$

16. Twice a number subtracted from 8.

    a. $8 - 2x$ **b.** $2x - 8$ **c.** $8x + 2$ **d.** $8x - 2$

Chapter 3 Practice Test B (cont'd)

Translate each sentence into an equation. Use x to represent "a number."

17. The sum of 2 times a number and 7 is 5.
    a. $2x = 7 + 5$ **b.** $7x + 2 = 5$ **c.** $2x + 7 = 5$ **d.** $2(x + 7) = 5$

18. Eight added to 3 times a number equals -12.
    a. $8x + 3 = -12$ **b.** $3(x + 8) = -12$ **c.** $3x = 8 - 12$ **d.** $3x + 8 = -12$

Solve.

19. Seven subtracted from twice a number is 11. Find the number.
    a. 2   **b.** 9   **c.** -2 **d.** 6

20. In a college algebra class, there are 6 more females than males. If there are 42 students in the class, how many males are there?
    a. 24   **b.** 18   **c.** 21   **d.** 6

4.1 Introduction to Fractions and Mixed Numbers

Learning Objectives
   A. Identify the numerator and denominator of a fraction.
   B. Write a fraction to represent parts of figures or real-life data.
   C. Graph fractions on a number line.
   D. Review division properties for 0 and 1.
   E. Write mixed numbers as improper fractions.
   F. Write improper fractions as mixed numbers or whole numbers.

Vocabulary
fraction, numerator, denominator, proper fraction, improper fraction, mixed number

---

Objective A Identify the numerator and denominator of a fraction.

   Identify the numerator and denominator of each fraction.

   Example 1 $\dfrac{3}{8}$

   Example 2 $\dfrac{7}{12z}$

Objective B Write a fraction to represent parts of figures or real-life data.

   Write a fraction to represent the shaded part of each figure.

   Example 3

   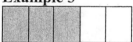

   Example 4

1) numerator = 3, denominator = 8, 2) numerator = 7, denominator = 12z, 3) $\dfrac{3}{5}$, 4) $\dfrac{3}{4}$

Examples 4.1 (cont'd)

Example 5

Example 6 The line represents what fraction of an inch?

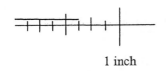

1 inch

Example 7 Shade $\dfrac{5}{6}$ of the figure shown.

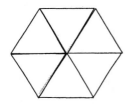

Example 8 Shade $\dfrac{2}{3}$ of the figure shown.

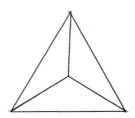

5) $\dfrac{3}{7}$, 6) $\dfrac{5}{8}$, 7) ⬡ , 8) ◭

Examples 4.1 (cont'd)

Objective B Write fractions from real life data.

Example 9 In a class of 32 students, 17 are seniors. What fraction of the class do the seniors represent?

Represent the shaded part of each figure as both an improper fraction and a mixed number.

Example 10

Example 11

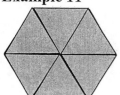 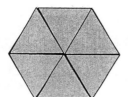 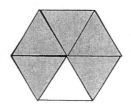

Objective C Graph fractions on a number line.

Example 12 Graph each proper fraction on a number line.

(a) $\dfrac{7}{8}$ (b) $\dfrac{1}{4}$ (c) $\dfrac{2}{3}$

0   1   0   1   0   1

Example 13 Graph each improper fraction on the number line.

(a) $\dfrac{5}{4}$ (b) $\dfrac{7}{4}$ (c) $\dfrac{5}{5}$ (d) $\dfrac{2}{1}$

0   1   2

9) $\dfrac{17}{32}$, 10) $\dfrac{3}{2}$, $1\dfrac{1}{2}$, 11) $\dfrac{17}{6}$, $2\dfrac{5}{6}$, 12a) ,b)

c)   , 13)

Examples 4.1 (cont'd)

Objective D Review division properties for 0 and 1.

Simplify.

Example 14 $\dfrac{10}{10}$

Example 15 $\dfrac{-8}{-8}$

Example 16 $\dfrac{0}{-2}$

Example 17 $\dfrac{-12}{1}$

Example 18 $\dfrac{82}{1}$

Example 19 $\dfrac{-9}{0}$

Objective D Write mixed numbers as improper fractions.

Example 20 Write each as an improper fraction.

(a) $3\dfrac{3}{8}$

(b) $1\dfrac{1}{19}$

Objective E Write improper fractions as mixed numbers or whole numbers.

Example 21 Write each as a mixed number or a whole number.

(a) $\dfrac{29}{3}$

(b) $\dfrac{13}{11}$

(c) $\dfrac{60}{5}$

14) 1, 15) 1, 16) 0, 17) −12, 18) 82, 19) undefined, 20a) $\dfrac{27}{8}$, b) $\dfrac{20}{19}$, 21a) $9\dfrac{2}{3}$, b) $1\dfrac{2}{11}$, c) 12

Practice Set 4.1

Use the choices below to fill in each blank.

improper	**fraction**	**proper**	**undefined**
mixed number	**denominator**	**numerator**	

1. The number $\dfrac{17}{25}$ is called a _____. The number 17 is called its
   _____ and 25 is called its _____.

2. The fraction $\dfrac{11}{27}$ is called a(n) _____ fraction. The fraction is $\dfrac{17}{8}$
   called a(n) _____ fraction.

3. The number $7\dfrac{2}{5}$ is a(n) _____ .

Objective A Identify the numerator and denominator of a fraction.

4. Use the fraction $\dfrac{18}{29}$ to identify the parts.

   4. numerator: _____

      denominator: _____

Objective B Write a fraction to represent parts of figures or real-life data.

Write a proper fraction to represent the shaded part of each figure.

5.

5. _____

6.

1 inch

6. _____

Practice Set 4.1 (cont'd)

Draw and shade a part of a figure to represent each fraction.

7. $\dfrac{3}{5}$

7. _____

8. $\dfrac{2}{3}$

8. _____

Write each fraction.

9. 271 students out of 519 at a large high school are seniors. What fraction of the students are seniors?

9. _____

10. In a leap year, there 29 days in February. Twelve days represents what fraction of February in a leap year?

10. _____

Objective C Graph fractions on a number.

Graph each fraction on a number line.

11. $\dfrac{3}{4}$

11.
     0   1

12. $\dfrac{2}{3}$

12. 0   1

Practice Set 4.1 (cont'd)

Objective D Review division properties for 0 and 1.

Simplify by dividing.

13. $\dfrac{-23}{-23}$ 13. _____

14. $\dfrac{7}{7}$ 14. _____

15. $\dfrac{0}{6}$ 15. _____

16. $\dfrac{-5}{0}$ 16. _____

Objective E Write mixed numbers as improper fractions.

Write each mixed number as an improper fraction.

17. $3\dfrac{5}{7}$ 17. _____

18. $5\dfrac{1}{11}$ 18. _____

19. $12\dfrac{2}{3}$ 19. _____

Practice Set 4.1 (cont'd)

Objective F Write improper fractions as mixed numbers or whole numbers.

Write each improper fraction as a mixed number or as a whole number.

20. $\dfrac{23}{7}$ 20. _____

21. $\dfrac{53}{10}$ 21. _____

22. $\dfrac{105}{104}$ 22. _____

23. $\dfrac{37}{37}$ 23. _____

Extensions

24. Write the fraction in two other equivalent ways by   24. _____
    inserting the negative sign in different places. $-\dfrac{3}{5}$

25. If the following represents $\dfrac{3}{7}$ of a diagram, draw   25. _____
    the whole diagram.

4.2 Factors and Simplest Form

Learning Objectives
  A. Write a number as a product of prime numbers.
  B. Write a fraction in simplest form.
  C. Determine whether two fractions are equivalent.
  D. Solve problems by writing fractions in simplest form.

Vocabulary
composite number, prime number

Objective A Write a number a product of prime numbers.

Example 1 Write the prime factorization of 12.

Example 2 Write the prime factorization of 60.

Example 3 Write the prime factorization of 360.

Objective B Write fractions in simplest form.

Example 4 Write in simplest form: $\dfrac{8}{12}$

Example 5 Write in simplest form: $\dfrac{15x}{25}$

1) $2^2 \cdot 3$, 2) $2^2 \cdot 3 \cdot 5$, 3) $2^3 \cdot 3^2 \cdot 5$, 4) $\dfrac{2}{3}$, 5) $\dfrac{3x}{5}$

Examples 4.2 (cont'd)

Example 6 Write in simplest form: $-\dfrac{8}{35}$

Example 7 Write in simplest form: $\dfrac{84}{504}$

Example 8 Write in simplest form: $\dfrac{-24}{42}$

Example 9 Write in simplest form: $\dfrac{15x^2}{35x^4}$

Objective C Determine whether two fractions are equivalent.

Example 10 Determine whether $\dfrac{15}{20}$ and $\dfrac{18}{24}$ are equivalent.

Example 11 Determine whether $\dfrac{7}{22}$ and $\dfrac{10}{30}$ are equivalent.

Objective D Solve problems by writing fractions in simplest form.

Example 12 There are 58 national parks in the United States. Eight of theses parks are in California. What fraction of the national parks are in California? Write the fraction in lowest form.

6) $-\dfrac{8}{35}$, 7) $\dfrac{1}{6}$, 8) $-\dfrac{4}{7}$, 9) $\dfrac{3}{7x^2}$, 10) yes, 11) no, 12) $\dfrac{4}{29}$

Practice Set 4.2

Objective A Write a number as a product of prime factors.

Write the prime factorization of each number.

1. 24

1. _____

2. 36

2. _____

3. 70

3. _____

4. 378

4. _____

Objective B Write a fraction in simplest form.

Write each fraction in lowest form.

5. $\dfrac{4}{18}$

5. _____

6. $\dfrac{9}{27}$

6. _____

7. $-\dfrac{10}{35}$

7. _____

8. $\dfrac{4x^2 y}{8x^3 y}$

8. _____

9. $\dfrac{35x}{45x^2 y}$

9. _____

Objective C Determine whether tow fractions are equivalent.
Determine whether each pair of fractions are equivalent.

10. $\dfrac{6}{15}$ and $\dfrac{8}{20}$

10. _____

11. $\dfrac{14}{21}$ and $\dfrac{4}{8}$

11. _____

12. $\dfrac{6}{14}$ and $\dfrac{9}{31}$

12. _____

Martin-Gay **Prealgebra** edition 5   119

Practice Set 4.2 (cont'd)

13. $\dfrac{2}{11}$ and $\dfrac{6}{21}$ 13. _____

14. $\dfrac{25}{30}$ and $\dfrac{10}{12}$ 14. _____

Objective D Solve problems by writing fractions in lowest terms.

Solve. Write each fraction in lowest terms.

15. Jose spends 6 hours of each week day at school.   15. _____
    If he is awake 18 hours, what fraction of his waking
    hours is he at school?

16. Sixteen states in the United States have over half of   16. _____
    their area in the Central Time Zone. What fraction
    of the 50 states are mostly in the Central Time Zone?

17. Oklahoma Memorial Stadium at the University   17. _____
    of Oklahoma holds 86,000 spectators (to the nearest
    thousand). Jones AT&T Stadium at Texas Tech
    University holds 56,000 spectators (to the nearest
    thousand). What fraction of Oklahoma spectators
    would fit in the Texas Tech stadium?

18. Eighteen out of 32 students in one class were   18. _____
    freshmen. What fraction of the class were
    freshmen?

Extensions

19. Write in lowest terms: $\dfrac{7560}{10080}$ 19. _____

20. Is 38,538 divisible by 18?   20. _____

4.3 Multiplying and Dividing Fractions

Learning Objectives
   A. Multiply fractions.
   B. Evaluate exponential expressions with fractional bases.
   C. Divide fractions.
   D. Multiply and divide given replacement values.
   E. Solve applications that require multiplication of fractions.

Vocabulary
reciprocal, of

Objective A Multiply fractions.

   Multiply.

Example 1 $\dfrac{1}{5} \cdot \dfrac{2}{3}$

Example 2 $\dfrac{3}{7} \cdot \dfrac{2}{5}$

Multiply and simplify.

Example 3 $\dfrac{2}{5} \cdot \dfrac{1}{4}$

Example 4 $\dfrac{-3}{5} \cdot \dfrac{5}{6}$

Example 5 $\dfrac{17}{21} \cdot \dfrac{7}{8}$

1) $\dfrac{2}{15}$,   2) $\dfrac{6}{35}$,   3) $\dfrac{1}{10}$,   4) $-\dfrac{1}{2}$,   5) $\dfrac{17}{24}$

Examples 4.3 (cont'd)

Example 6 $\left(-\dfrac{3}{20}\right)\left(-\dfrac{4}{9}\right)$

Example 7 $\dfrac{1}{6} \cdot \dfrac{2}{5} \cdot \dfrac{3}{4}$

Example 8 $\dfrac{5x}{11} \cdot \dfrac{22}{25x}$

Example 9 $\dfrac{x^3}{y^2} \cdot \dfrac{y^4}{x^2}$

Example 10 Evaluate.

(a) $\left(\dfrac{3}{4}\right)^3$ (b) $\left(-\dfrac{1}{5}\right)^4$

Objective C Divide fractions.

Divide and simplify.

Example 11 $\dfrac{4}{5} \div \dfrac{16}{15}$

6) $\dfrac{1}{15}$, *7)* $\dfrac{1}{20}$, *8)* $\dfrac{2}{5}$, *9)* xy^2, *10a)* $\dfrac{27}{64}$, *b)* $\dfrac{1}{625}$, *11)* $\dfrac{3}{4}$

Examples 4.3 (cont'd)

Example 12 $\dfrac{7}{10} \div \dfrac{1}{80}$

Example 13 $-\dfrac{21}{36} \div -\dfrac{7}{8}$

Example 14 $\dfrac{4x}{5} \div 16x^2$

Example 15 $\left(\dfrac{2}{3} \cdot \dfrac{3}{4}\right) \div \dfrac{1}{2}$

Objective D Multiply and divide given fractional replacement values.

Example 16 If $x = \dfrac{7}{15}$ and $y = -\dfrac{5}{9}$, evaluate (a) xy and (b) $x \div y$.

Example 17 Is $-\dfrac{1}{4}$ a solution of the equation $-\dfrac{2}{3}x = \dfrac{1}{6}$?

Objective E Solve applications that require multiplication of fractions.

Example 18 $\dfrac{2}{3}$ of a class made a C or better on the first test. If there are 27 in the class, how many students made a C or better?

12) 56, 13) $\dfrac{2}{3}$, 14) $\dfrac{1}{20x}$, 15) 1, 16a) $-\dfrac{7}{27}$, b) $-\dfrac{21}{25}$, 17) yes, 18) 18 students

Practice Set 4.3

Use the choices below to fill in each blank.

multiplication   **reciprocals**

1. The word "of" indicates _____.

2. To divide fractions, multiply by the _____ of the divisor.

Objective A Multiply fractions.

Multiply and simplify.

3. $\dfrac{3}{5} \cdot \dfrac{2}{7}$

3. _____

4. $\dfrac{-2}{5} \cdot 10x^2$

4. _____

5. $-\dfrac{4}{5} \cdot \dfrac{3}{8}$

5. _____

6. $\dfrac{4}{7} \cdot 0$

6. _____

7. $\dfrac{3x}{5y} \cdot \dfrac{5y}{6x^2}$

7. _____

Objective B Evaluate exponential expression with fractional bases.

8. $\left(\dfrac{1}{3}\right)^2$

8. _____

9. $\left(-\dfrac{1}{2}\right)^3$

9. _____

Practice Set 4.3 (cont'd)

Objective C Divide fractions.

10. $\dfrac{2}{3} \div \dfrac{1}{6}$

10. _____

11. $\dfrac{3}{5} \div \dfrac{3}{4x}$

11. _____

12. $\dfrac{8x}{15} \div \dfrac{5}{12}$

12. _____

Perform each operation.

13. $\dfrac{-2}{3} \cdot -\dfrac{5}{6}$

13. _____

14. $\dfrac{1}{2}\left(\dfrac{2}{3}\right)^{3}$

14. _____

15. $\left(\dfrac{3}{4} \div 6\right) \cdot \dfrac{1}{2}$

15. _____

16. $\dfrac{a^{3}b^{2}}{c} \div \dfrac{a^{5}b}{c^{5}}$

16. _____

17. $\dfrac{1}{2} \cdot \left(\dfrac{1}{3} \div \dfrac{1}{2}\right)$

17. _____

Practice Set 4.3 (cont'd)

Objective D Multiply and divide given fractional replacement values.

18. If $x = \dfrac{2}{3}$ and $y = \dfrac{2}{5}$, find xy and $x \div y$.

18. _____

Objective E Solve applications that require multiplication of fractions.

19. Find $\dfrac{2}{3}$ of 60.

19. _____

20. Find $\dfrac{5}{8}$ of 40.

20. _____

21. A patient was told that no more than $\dfrac{1}{5}$ of his calories should come from fat. If his diet consists of 1500 calories, how many of these calories can come from fat?

21. _____

22. The Franklin family sold a house for $75,000. If $\dfrac{7}{100}$ went to the realtor, how much did the realtor make?

22. _____

23. Find each answer. Are the two answers different? $\dfrac{3}{10} \div \dfrac{9}{25} \cdot \dfrac{5}{6}$ and $\dfrac{3}{10} \div \left(\dfrac{9}{25} \cdot \dfrac{5}{6} \right)$

23. _____

24. Marianne bought a blouse on sale for $\dfrac{1}{3}$ off. If the original price was $48, what was the sale price?

24. _____

Martin-Gay **Prealgebra** edition 5

4.4 Adding and Subtracting Like Fractions, Least Common Denominator, and Equivalent Fractions

Learning Objectives
- A. Add or subtract like fractions.
- B. Add or subtract given fractional replacement values.
- C. Solve problems by adding or subtracting like fractions.
- D. Find the least common denominator of a list of fractions.
- E. Write equivalent fractions

Vocabulary
like fractions, unlike fractions

Objective A Add or subtract like fractions.

Add and simplify.

Example 1 $\dfrac{3}{5}+\dfrac{1}{5}$

Example 2 $\dfrac{3}{8}+\dfrac{1}{8}$

Example 3 $\dfrac{4}{15}+\dfrac{7}{15}+\dfrac{4}{15}$

Subtract and simplify.

Example 4 $\dfrac{7}{11}-\dfrac{4}{11}$

Example 5 $\dfrac{5}{12}-\dfrac{1}{12}$

Example 6 $-\dfrac{7}{8}+\dfrac{3}{8}$

1) $\dfrac{4}{5}$, 2) $\dfrac{1}{2}$, 3) 1, 4) $\dfrac{3}{11}$, 5) $\dfrac{1}{3}$, 6) $-\dfrac{1}{2}$

Examples 4.4 (cont'd)

Example 7 Subtract. $\dfrac{3x}{5} - \dfrac{2}{5}$

Example 8 Subtract. $\dfrac{8}{15} - \dfrac{2}{15} - \dfrac{11}{15}$

Objective B Add and subtract given fractional replacement values.

Example 9 Evaluate $x - y$ if $x = \dfrac{4}{5}$ and $y = -\dfrac{3}{5}$.

Objective C Solve problems by adding or subtracting like fractions.

Example 10 Find the perimeter of the triangle.

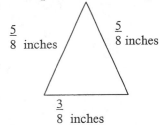

$\dfrac{5}{8}$ inches $\dfrac{5}{8}$ inches

$\dfrac{3}{8}$ inches

Example 11 The distance from home to work is $\dfrac{3}{4}$ mile. The distance from home to school is $\dfrac{1}{4}$ mile. How much further is it from home to work than from home to school?

Objective D Find the least common denominator of a list of fractions.

Example 12 Find the LCD of $\dfrac{3}{5}$ and $\dfrac{7}{10}$.

7) $\dfrac{3x-2}{5}$, 8) $-\dfrac{1}{3}$, 9) $1\dfrac{2}{5}$, 10) $1\dfrac{5}{8}$ *inch*, 11) $\dfrac{1}{2}$ *mile*, 12) 10

Examples 4.4 (cont'd)

Example 13 Find the LCD of $\dfrac{7}{15}$ and $\dfrac{5}{12}$.

Example 14 Find the LCD of $\dfrac{9}{36}$ and $\dfrac{7}{45}$.

Example 15 Find the LCD of $\dfrac{7}{8}$, $\dfrac{5}{12}$, and $\dfrac{7}{18}$.

Example 16 Find the LCD of $\dfrac{2}{7}$, $\dfrac{1}{x}$, and $\dfrac{4}{x^2}$.

Objective E Write equivalent fractions.

Example 17 Write $\dfrac{5}{7}$ as an equivalent fraction with a denominator of 21.

Example 18 Write an equivalent fractions with the given denominator. $\dfrac{3}{16} = \dfrac{}{48}$

Example 19 Write an equivalent fractions with the given denominator. $\dfrac{4x}{5} = \dfrac{}{45}$

Example 20 Write an equivalent fractions with the given denominator. $4 = \dfrac{}{12}$

Example 21 Write an equivalent fractions with the given denominator. $\dfrac{3}{16x} = \dfrac{}{48x}$

13) 60, 14) 180, 15) 72, 16) $7x^2$, 17) $\dfrac{15}{21}$, 18) $\dfrac{9}{48}$, 19) $\dfrac{36x}{45}$, 20) $\dfrac{48}{12}$, 21) $\dfrac{9}{48x}$

Practice Set 4.4

Use the choices below to fill in each blank.

least common denominator   **like**
unlike   **equivalent**

1. The fractions $\frac{3}{7}$ and $\frac{4}{3}$ are called _____ fractions while the

   fractions $\frac{5}{6}$ and $\frac{1}{6}$ are called _____ fractions.

2. The smallest positive number divisible by all the denominators of a list of fractions is
   called the _____.

3. Fractions that can be reduced to the same value are called _____.

Objective A Add or subtract like fractions.

Perform the indicated operations. Simplify all fractions.

4. $\frac{2}{5} + \frac{1}{5}$

4. _____

5. $\frac{7}{12} + \frac{7}{12}$

5. _____

6. $\frac{3}{4x} + \frac{2}{4x}$

6. _____

7. $\frac{8}{15} - \frac{7}{15}$

7. _____

8. $\frac{5}{x} - \frac{3}{x}$

8. _____

9. $\frac{7}{10} - \frac{5}{10}$

9. _____

Practice Set 4.4 (cont'd)

10. $\dfrac{1}{6} - \dfrac{4}{6} + \dfrac{2}{6}$

10. _____

11. $\dfrac{4x}{3} + \dfrac{1}{3}$

11. _____

Objective B Add or subtract given fractional replacement values.

12. Evaluate $x + y$ for $x = \dfrac{3}{5}$ and $y = \dfrac{1}{5}$.

12. _____

13. Evaluate $x - y$ for $\dfrac{2}{3}$ and $y = -\dfrac{1}{3}$.

13. _____

Objective C Solve problems by adding or subtracting like fractions.

14. Find the perimeter of the triangle.

14. _____

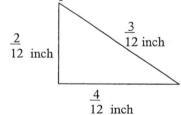

$\dfrac{2}{12}$ inch

$\dfrac{3}{12}$ inch

$\dfrac{4}{12}$ inch

15. Find the perimeter of the rectangle.

15. _____

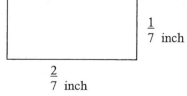

$\dfrac{1}{7}$ inch

$\dfrac{2}{7}$ inch

Practice Set 4.4 (cont'd)

16. Amy ran $\frac{4}{3}$ of a mile away from her home and

then turned toward home and ran $\frac{2}{3}$ of a mile

before she stopped to rest. How far from home was
she when she stopped?

16. _____

Objective D Find the least common denominator of a list of fractions.

Find the LCD for each list of fractions.

17. $\frac{2}{3}, \frac{1}{5}$

17. _____

18. $\frac{5}{6}, \frac{1}{4}$

18. _____

19. $\frac{1}{5}, \frac{1}{15}, \frac{1}{6}$

19. _____

Objective E Write equivalent fractions.

Write each fraction as an equivalent fraction with the given denominator.

20. $\frac{4}{5} = \frac{}{20}$

20. _____

21. $\frac{4}{7} = \frac{}{28}$

21. _____

22. $\frac{2x}{6} = \frac{}{24}$

22. _____

Extensions

Evaluate.

23. $2 \cdot \left(\frac{1}{4}\right)^2$

23. _____

24. $\frac{1}{4} \cdot \left(-\frac{2}{3}\right)^3 + \frac{1}{6}$

24. _____

Name:

Instructor:

Date:

Section:

4.5 Adding and Subtracting Unlike Fractions

Learning Objectives
    A. Add or subtract unlike fractions.
    B. Write fractions in order.
    C. Evaluate expressions given fractional replacement values.
    D. Solve problems by adding or subtracting unlike fractions.

Objective A Add or subtract unlike fractions.

Example 1 Add: $\dfrac{2}{3}+\dfrac{1}{6}$

Example 2 Add: $\dfrac{2x}{15}+\dfrac{3x}{10}$

Example 3 Add: $-\dfrac{1}{4}+\dfrac{1}{5}$

Example 4 Subtract: $\dfrac{3}{7}-\dfrac{7}{10}$

Example 5 Find: $-\dfrac{2}{5}+\dfrac{1}{3}-\dfrac{5}{6}$

Example 6 Subtract: $4-\dfrac{x}{2}$

1) $\dfrac{5}{6}$, 2) $\dfrac{13x}{30}$, 3) $-\dfrac{1}{20}$, 4) $-\dfrac{19}{70}$, 5) $-\dfrac{9}{10}$, 6) $\dfrac{8-x}{2}$

Martin-Gay **Prealgebra** edition 5

133

Example 4.5 (cont'd)

Objective B Write fractions in order.

Example 7 Insert < or > to make a true sentence.

$\dfrac{3}{5}$ ___ $\dfrac{4}{27}$

Example 8 Insert < or > to make a true sentence.

$-\dfrac{2}{7}$ ___ $-\dfrac{8}{23}$

Objective C Evaluate expressions given fractional replacement values.

Example 9 Evaluate $x - y$ if $x = \dfrac{3}{5}$ and $y = \dfrac{4}{9}$.

Objective D Solve problems by adding or subtracting unlike fractions.

Example 10 Jan buys $\dfrac{1}{4}$ yard of red material, $\dfrac{1}{8}$ yard of white material, and $\dfrac{2}{5}$ yard of blue material. Find the total number of yards of material she purchased.

Example 11 Payton drives $\dfrac{1}{2}$ hour to school everyday. If he has driven $\dfrac{1}{3}$ hour, what fraction of an hour does he have left to go?

7) >, 8) >, 9) $\dfrac{7}{45}$, 10) $\dfrac{31}{40}$ yard, 11) $\dfrac{1}{6}$ hour

Practice Set 4.5

Objective A Add or subtract unlike fractions.

Add or subtract as indicated.

1. $\dfrac{1}{4}+\dfrac{3}{5}$

1. _____

2. $\dfrac{1}{12}+\dfrac{2}{3}$

2. _____

3. $\dfrac{2y}{3}+\dfrac{5y}{21}$

3. _____

4. $3-\dfrac{2}{3}$

4. _____

5. $\dfrac{4b}{3}-\dfrac{1}{6}$

5. _____

6. $\dfrac{3}{8}-\dfrac{1}{4}$

6. _____

7. $-\dfrac{2}{5}-\dfrac{1}{4}$

7. _____

8. $\dfrac{3}{13}+\dfrac{1}{5}$

8. _____

9. $-\dfrac{2}{3}-\dfrac{1}{7}$

9. _____

10. $\dfrac{4x}{2}-\dfrac{1}{8}$

10. _____

11. $-\dfrac{3}{5}+\dfrac{2}{3}-\dfrac{4}{15}$

11. _____

Practice Set 4.5 (cont'd)

Objective B Write fractions in order.

12. Insert < or > to form a true sentence. $\dfrac{1}{3}$ $\dfrac{4}{13}$ 12. _____

13. Insert < or > to form a true sentence. $-\dfrac{3}{4}$ $-\dfrac{11}{15}$ 13. _____

Objective C Evaluate expressions given fractions values.

Evaluate each expression if $x = -\dfrac{1}{4}$ and $y = \dfrac{2}{3}$.

14. $x + y$ 14. _____

15. $x - y$ 15. _____

Practice Set 4.5 (cont'd)

16. $x \div y$ 16. _____

17. $2x - y$ 17. _____

Objective D Solve problems by adding and subtracting unlike fractions.

18. Find the perimeter of the parallelogram.   18. _____

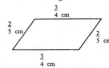

19. Find the inner diameter of the washer.   19. _____

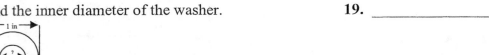

Extensions

20. In one basketball game, the Edwards High School   20. _____
    team scores $\dfrac{2}{7}$ of the total points for both teams by

    field goals. They scored $\dfrac{1}{5}$ of the total points by

    free shots. What fraction of the total points did
    Edwards High School score? Did they win the game?

4.6 Complex Fractions and Review of Order of Operations

Learning Objectives
    A. Simplify complex fractions.
    B. Review order of operations.
    C. Evaluate expressions given replacement values.

Vocabulary
complex fraction

Objective A Simplify complex fractions.

Simplify.

Example 1
$$\dfrac{\dfrac{x}{3}}{\dfrac{5}{6}}$$

Example 2
$$\dfrac{\dfrac{2}{3}+\dfrac{1}{2}}{\dfrac{3}{4}-\dfrac{1}{3}}$$

Example 3
$$\dfrac{\dfrac{2}{3}+\dfrac{1}{4}}{\dfrac{5}{6}-\dfrac{1}{4}}$$

1) $\dfrac{2x}{5}$, *2)* $\dfrac{14}{5}$, *3)* $1\dfrac{4}{7}$

Examples 4.6 (cont'd)

Example 4 $\dfrac{\dfrac{x}{2}-1}{\dfrac{2}{5}}$

Objective B Review the order of operations.

Example 5 Simplify: $2-\left(\dfrac{3}{4}\right)^2$

Example 6 Simplify: $\dfrac{1}{3}-\dfrac{2}{5}\left(\dfrac{1}{3}-\dfrac{2}{5}\right)$

Objective C Evaluate expression given replacement values.

Example 7 Evaluate $2x-3y$ for $x=-\dfrac{3}{4}$ and $y=\dfrac{1}{5}$.

4) $\dfrac{5x-10}{4}$, 5) $1\dfrac{7}{16}$, 6) $\dfrac{9}{25}$, 7) $-2\dfrac{1}{10}$

Name: Date:

Instructor: Section:

Practice Set 4.6

For each pair of fractions, indicate the complex fraction.

1. $\dfrac{\frac{1}{2}}{3}$ $\dfrac{17}{24}$

1. _____

2. $\dfrac{3x-2}{14}$ $\dfrac{2}{\frac{1}{4}}$

2. _____

Objective A Simplify complex fractions.

3. $\dfrac{\frac{2}{5}}{\frac{3}{10}}$

3. _____

4. $\dfrac{\frac{3}{7}}{\frac{2}{7}}$

4. _____

5. $\dfrac{\frac{3x}{4}}{\frac{7}{8}}$

5. _____

6. $\dfrac{\frac{1}{3}+\frac{2}{5}}{\frac{5}{6}}$

6. _____

7. $\dfrac{1+\frac{2}{3}}{\frac{5}{6}}$

7. _____

8. $\dfrac{\frac{1}{2x}}{1-\frac{1}{4}}$

8. _____

Practice Set 4.6 (cont'd)

Objective B Review the order of operations.

Use the order of operations to simplify each expression.

9. $\dfrac{1}{2} + \dfrac{1}{3} \cdot \dfrac{1}{2}$

9. _____

10. $\dfrac{2}{3} \div \dfrac{4}{5} \cdot \dfrac{3}{5}$

10. _____

11. $\left(\dfrac{3}{5} + \dfrac{1}{3} \right)\left(\dfrac{1}{15} - \dfrac{1}{3} \right)$

11. _____

12. $\left(\dfrac{1}{2} \right)^2 - \left(\dfrac{1}{3} \right)^2$

12. _____

13. $\left(\dfrac{1}{2} - \dfrac{1}{3} \right)^2$

13. _____

14. $\dfrac{3}{4} \div \left(\dfrac{1}{2} - \dfrac{1}{6} \right)$

14. _____

15. $\left(1 - \dfrac{2}{3} \right)^2$

15. _____

Objective C Evaluate expressions given replacement values.

Evaluate each for $x = \dfrac{2}{3}$ and $y = -\dfrac{1}{4}$.

16. $2x + y$

16. _____

17. $x^2 - y$

17. _____

18. $\dfrac{x + y}{2}$

18. _____

19. $(1 - x)(1 + x)$

19. _____

Concept Extensions

20. Find the average. $\dfrac{2}{5}, \dfrac{4}{15}$

20. _____

4.7 Operations on Mixed Numbers

Learning Objectives
 A. Graph positive and negative fractions and mixed numbers.
 B. Multiply or divide mixed or whole numbers.
 C. Add or subtract whole numbers.
 D. Solve problems containing mixed numbers.
 E. Perform operations on negative mixed numbers.

Objective A Graph positive and negative fractions and mixed numbers.

Example 1 Graph the numbers on a number line.

$$-2\frac{1}{2},\ -1,\ 1\frac{1}{3},\ 2$$

Objective B Multiply or divide mixed numbers or whole numbers.

Example 2 Multiply: $2\frac{1}{2} \cdot \frac{3}{5}$

Example 3 Multiply: $\frac{1}{5} \cdot 15$

Example 4 Mulptiply: $1\frac{2}{5} \cdot 2\frac{1}{4}$ Check by estimating.

Example 5 Multiply: $6 \cdot 3\frac{1}{9}$ Check by estimating.

Divide.

Example 6 $\frac{2}{4} \div 3$

1)   *, 2)* $1\frac{1}{2}$, *3)* 3, *4)* $3\frac{3}{10}$, *5)* $18\frac{2}{3}$, *6)* $\frac{1}{6}$

Examples 4.7 (cont'd)

Example 7 $\dfrac{11}{12} \div 1\dfrac{1}{3}$

Example 8 $4\dfrac{1}{5} \cdot 2\dfrac{1}{3}$

Objective C   Add or subtract mixed numbers.

Example 9   Add: $1\dfrac{1}{5} + 3\dfrac{1}{4}$ Check by estimating.

Example 10   Add: $2\dfrac{3}{4} + 1\dfrac{7}{12}$

Example 11   Add: $2\dfrac{7}{8} + 4 + 1\dfrac{3}{4}$

Example 12   Subtract: $3\dfrac{3}{5} - 1\dfrac{1}{4}$

Example 13   Subtract: $6\dfrac{3}{5} - 4\dfrac{11}{15}$

Example 14   Subtract: $6 - 4\dfrac{2}{7}$

7) $\dfrac{11}{16}$,   8) $9\dfrac{4}{5}$,   9) $4\dfrac{9}{20}$,   10) $4\dfrac{1}{3}$,   11) $8\dfrac{5}{8}$,   12) $2\dfrac{7}{20}$,   13) $1\dfrac{13}{15}$,   14) $1\dfrac{5}{7}$

Examples 4.7 (cont'd)

Objective D Solve problems containing mixed numbers.

Example 15 Larry caught a trout that was $8\frac{3}{4}$ inches. Jeff caught a trout that was $9\frac{1}{16}$ inches. How much longer was Jeff's fish than Larry's?

Example 16 There is a 60 inch board that needs to be cut into pieces that are $3\frac{3}{4}$ in long. How many pieces can be made?

Objective E Perform operations on negative mixed numbers.

Write each as a fraction.

Example 17 $-3\frac{7}{11}$

Example 18 $-17\frac{1}{5}$

Write each as a mixed number.

Example 19 $-\frac{37}{5}$

Example 20 $-\frac{10}{7}$

15) $\frac{5}{16}$ inch, 16) 16 pieces, 17) $-\frac{40}{11}$, 18) $-\frac{86}{5}$, 19) $-7\frac{2}{5}$, 20) $-1\frac{3}{7}$

Examples 4.7 (cont'd)

Perform the indicated operations.

Example 21 $-2\dfrac{1}{4} \cdot 3\dfrac{1}{5}$

Example 22 $-2\dfrac{4}{7} \div \left(-3\dfrac{1}{5}\right)$

Example 23 Add: $7\dfrac{2}{3} + \left(-8\dfrac{1}{5}\right)$

Example 24 Subtract: $4\dfrac{8}{9} - \left(-3\dfrac{2}{9}\right)$

21) $-7\dfrac{1}{5}$, *22)* $\dfrac{45}{56}$, *23)* $-\dfrac{8}{15}$, *24)* $8\dfrac{1}{9}$

Practice Set 4.7

Use the choices below to fill in each blank.

fraction	improper	mixed number
round	whole number	

1. The number $-2\dfrac{1}{5}$ is called a(n) _____.

2. For the number $8\dfrac{3}{7}$, the number 8 is called the _____ and $\dfrac{3}{7}$ is called the _____ part.

3. The number $\dfrac{47}{3}$ is a(n) _____ fraction.

Objective A Graph positive and negative fractions and mixed numbers.

4. Graph each on the numbers on the number line.

   $-1,\ -1\dfrac{3}{4},\ 0,\ 2,\ 1\dfrac{1}{2}$

4.

Objective B Multiply or divide mixed or whole numbers.

5. $3\dfrac{1}{2}\cdot\dfrac{1}{5}$

5. _____

6. $2\dfrac{1}{5}\cdot3\dfrac{4}{7}$

6. _____

7. $8\cdot3\dfrac{1}{4}$

7. _____

8. $\dfrac{4}{9}\div2\dfrac{1}{5}$

8. _____

9. $2\dfrac{1}{5}\div3\dfrac{1}{5}$

9. _____

Practice Set 4.7 (cont'd)

Objective C Add or subtract mixed numbers.

10. $3\frac{1}{7} + 4\frac{2}{3}$ 10. _____

11. $9\frac{3}{7} + 4\frac{2}{3}$ 11. _____

12. $5\frac{4}{7} - 2\frac{1}{3}$ 12. _____

13. $7\frac{1}{2} - 4\frac{3}{4}$ 13. _____

14. $3 - 2\frac{3}{5}$ 14. _____

Objective D Solve problems containing mixed numbers.

15. Find the area of a porch in the shape of a rectangle   15. _____

     that is $7\frac{1}{2}$ feet by $3\frac{1}{4}$ feet.

Objective E Perform operations on negative mixed numbers.

16. $-3\frac{1}{5} \cdot 2\frac{1}{7}$ 16. _____

17. $-2\frac{1}{4}\left(-4\frac{1}{3}\right)$ 17. _____

18. $7\frac{3}{8} - 9\frac{1}{4}$ 18. _____

19. $-4\frac{1}{5} - 2\frac{4}{7}$ 19. _____

Extensions

20. Insert < or > to make a true statement. $-\frac{3}{5}$ $-\frac{7}{10}$ 20. _____

4.8 Solving Equations Containing Fractions

Learning Objectives
  A. Solve equations containing fractions.
  B. Solve equations by multiplying by the LCD.
  C. Review adding and subtracting fractions.

Objective A Solve equations containing fractions.

Solve.

Example 1 $x - \dfrac{1}{5} = \dfrac{3}{4}$

Example 2 $\dfrac{2}{3}x = 8$

Example 3 $-\dfrac{2}{3}a = 12$

Example 4 $-\dfrac{4}{5}x = \dfrac{7}{10}$

Example 5 $6y = -\dfrac{6}{7}$

1) $\dfrac{19}{20}$, 2) 12, 3) –18, 4) $-\dfrac{7}{8}$, 5) $-\dfrac{1}{7}$

Examples 4.8 (cont'd)

Objective B Solve equations by multiplying by the LCD.

Solve

Example 6 $\dfrac{7}{8}x = \dfrac{1}{16}$

Example 7 $\dfrac{x}{2} - x = \dfrac{1}{5}$

Example 8 $\dfrac{x}{4} - \dfrac{x}{5} = 4$

Example 9 $\dfrac{x}{5} = \dfrac{x}{3} - 2$

Objective C Review of adding and subtracting fractions.

Example 10 Add: $\dfrac{x}{4} + \dfrac{3}{8}$

6) $\dfrac{1}{14}$, 7) $-\dfrac{2}{5}$, 8) 80, 9) 15, 10) $\dfrac{2x+3}{8}$

Practice Set 4.8

Objective A Solve equations containing fractions.

Solve each equation.

1. $x - \dfrac{1}{5} = \dfrac{3}{4}$

1. _____

2. $x + \dfrac{2}{3} = \dfrac{4}{5}$

2. _____

3. $3x - \dfrac{2}{5} = 2x + \dfrac{1}{7}$

3. _____

4. $6x = 5$

4. _____

5. $-2x = 1$

5. _____

6. $\dfrac{3}{5}x = -4$

6. _____

Practice Set 4.8 (cont'd)

Objective B Solve equations by multiplying by the LCD.

7. $\dfrac{3}{7}x = \dfrac{3}{14}$ 7. _____

8. $\dfrac{x}{5} - x = -4$ 8. _____

9. $\dfrac{x}{2} - 5 = \dfrac{1}{4}$ 9. _____

10. $\dfrac{3}{5} - \dfrac{1}{4} = \dfrac{x}{20}$ 10. _____

11. $\dfrac{7}{8}x = \dfrac{7}{12}$ 11. _____

12. $\dfrac{x}{3} - 1 = \dfrac{2}{3}$ 12. _____

13. $-\dfrac{2x}{3} = -6$ 13. _____

Practice Set 4.8 (cont'd)

14. $17x - 14x = \dfrac{5}{6}$

14. _____

Objective C Review adding and subtracting fractions.

15. $\dfrac{x}{2} - \dfrac{2}{3}$

15. _____

16. $-\dfrac{4}{5} + \dfrac{y}{2}$

16. _____

17. $3 + \dfrac{5x}{3}$

17. _____

18. $\dfrac{2x}{5} + \dfrac{3x}{10}$

18. _____

Extensions

19. The area of the rectangle is $2\dfrac{1}{3}$ inches. Find its length x.

19. _____

x

$\dfrac{2}{3}$ inch

Chapter 4 Vocabulary Reference Sheet

Term	Definition	Example
Section 4.1		
Fraction	Part of a whole	$\frac{2}{3}$ is a fraction.
Numerator	Number of parts being considered.	In the fraction $\frac{2}{3}$, 2 is the numerator.
Denominator	Number of equal parts in the whole.	In the fraction $\frac{2}{3}$, 3 is the denominator.
Proper fraction	A fraction where the numerator is smaller than the denominator.	$\frac{2}{3}$ is a proper fraction.
Improper fraction	A fraction where the numerator is larger than the denominator.	$\frac{5}{3}$ is an improper fraction.
Mixed number	A whole number and a fraction.	$4\frac{2}{3}$ is a mixed number.
Section 4.2		
Prime number	A natural number that has exactly two different factors, 1 and itself.	7 is a prime number.
Section 4.3		
Reciprocal	The reciprocal of a number multiplied by the number is 1. To find the reciprocal of a number, interchange its numerator and denominator.	$\frac{2}{3}$ and $\frac{3}{2}$ are reciprocals.
Section 4.4		
Like fractions	Fractions that have the same denominator.	$\frac{2}{3}$ and $\frac{1}{3}$ are like fractions.
Unlike fractions	Fractions that have different denominators.	$\frac{2}{3}$ and $\frac{2}{5}$ are unlike fractions.
Least Common Denominator (LCD)	The smallest positive number divisible by all the denominators in a list of fractions.	For $\frac{1}{3}$ and $\frac{2}{5}$, 15 would be the LCD
Section 4.6		
Complex fraction	A fraction whose numerator or denominator or both contain fractions.	$\dfrac{\frac{2}{3}}{\frac{1}{5}+x}$ is a complex fraction.

Chapter 4 Practice Test A

1. Write a fraction to represent the shaded area.

    1. _____

2. Write the mixed number as an improper fraction. $6\frac{1}{3}$

    2. _____

3. Write the improper fraction as a mixed number. $\frac{28}{9}$

    3. _____

4. Write the fraction in simplest form. $\frac{32}{40}$

    4. _____

5. Write the fraction in simplest form. $\frac{25x}{40}$

    5. _____

6. Determine if the fractions are equivalent. $\frac{6}{32}$ and $\frac{9}{48}$

    6. _____

7. Find the prime factorization. 96

    7. _____

Perform each operation and write the answers in simplest form.

8. $\frac{2}{5} \div \frac{4}{5}$

    8. _____

9. $-\frac{2}{3} \cdot \frac{5}{6}$

    9. _____

10. $\frac{3x}{7} + \frac{x}{4}$

    10. _____

11. $\frac{2}{5} - \frac{3}{x}$

    11. _____

Chapter 4 Practice Test A (cont'd)

12. $-\dfrac{3}{7} \cdot \left(-\dfrac{5}{6}\right)$

12. _____

13. $-\dfrac{3}{5} \cdot \dfrac{5}{9}$

13. _____

14. $\dfrac{3a}{2} \cdot \dfrac{1}{6a}$

14. _____

15. $-\dfrac{3}{5} - \dfrac{1}{7}$

15. _____

16. $\dfrac{3a}{7} \div \dfrac{21a}{14}$

16. _____

17. $\dfrac{1}{12} - \dfrac{3}{4} + \dfrac{2}{3}$

17. _____

18. $6 - 3\dfrac{4}{5}$

18. _____

19. $2\dfrac{1}{5} \cdot 1\dfrac{2}{3}$

19. _____

Chapter 4 Practice Test A (cont'd)

20. $4\dfrac{1}{7} - 2\dfrac{2}{3}$

20. _____

21. $\dfrac{1}{5}\left(\dfrac{1}{3} - \dfrac{1}{2}\right)$

21. _____

22. $\left(\dfrac{1}{2}\right)^2 - \left(\dfrac{2}{3}\right)^2$

22. _____

Simplify each complex fraction.

23. $\dfrac{\dfrac{3}{5}}{\dfrac{9}{10}}$

23. _____

24. $\dfrac{2 - \dfrac{3}{5}}{\dfrac{4}{5} - \dfrac{1}{3}}$

24. _____

Solve each equation.

25. $-\dfrac{2}{3}x = \dfrac{6}{7}$

25. _____

Chapter 4 Practice Test A (cont'd)

26. $x - \dfrac{3}{5} = \dfrac{4}{3}$

26. _____

27. $\dfrac{x}{3} + \dfrac{1}{2} = \dfrac{2}{3} + x$

27. _____

Evaluate each expression for the given replacement values.

28. $x - 2y,\ \ x = \dfrac{1}{2}$ and $y = \dfrac{1}{4}$

28. _____

29. $x \div y,\ \ x = -\dfrac{2}{3}$ and $y = -\dfrac{1}{2}$

29. _____

30. A carpenter cuts a piece $3\dfrac{1}{2}$ feet off of a board that was 6 feet long. How much is left?

30. _____

Use the rectangle for problems 31 and 32.

$\dfrac{1}{3}$ foot

$\dfrac{3}{4}$ foot

31. Find the perimeter.

31. _____

32. Find the area.

32. _____

33. Ralph wants to cut a 60 inch board into pieces that were $7\dfrac{1}{2}$ inches. How many pieces could he cut?

33. _____

Chapter 4 Practice Test B

1. Write a fraction to represent the shaded area.

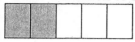

    a. $\dfrac{1}{5}$ **b.** $\dfrac{2}{5}$ **c.** $\dfrac{3}{5}$ **d.** $\dfrac{4}{5}$

2. Write the mixed number as an improper fraction. $5\dfrac{5}{6}$

    a. $\dfrac{55}{6}$ **b.** $\dfrac{30}{6}$ **c.** $\dfrac{35}{6}$ **d.** $\dfrac{60}{6}$

3. Write the improper fraction as a mixed number. $\dfrac{17}{6}$

    a. $2\dfrac{1}{6}$ **b.** $1\dfrac{11}{6}$ **c.** $3\dfrac{1}{6}$ **d.** $2\dfrac{5}{6}$

4. Write the fraction in simplest form. $\dfrac{9}{12}$

    a. $\dfrac{3}{4}$ **b.** $\dfrac{1}{3}$ **c.** $\dfrac{2}{3}$ **d.** $\dfrac{1}{4}$

5. Write the fractions in simplest form. $\dfrac{8}{12x}$

    a. $\dfrac{2x}{3}$ **b.** $\dfrac{2}{3}x$ **c.** $\dfrac{2}{3x}$ **d.** $\dfrac{4}{6x}$

6. Determine if the fractions are equivalent. $\dfrac{15}{25}$ $\dfrac{22}{36}$

    a. Yes   **b.** No

7. Find the prime factorization. 252
    a. $2\cdot3\cdot6\cdot7$ **b.** $2^2\cdot3\cdot7$ **c.** $2^2\cdot3^2\cdot7$ **d.** $2^2\cdot3^2\cdot7^2$

Perform the indicated operation and write the answer in simplest form.

8. $\dfrac{5}{12}\div\dfrac{5}{8}$

    a. $\dfrac{25}{96}$ **b.** $\dfrac{8}{12}$ **c.** $\dfrac{1}{3}$ **d.** $\dfrac{2}{3}$

9. $-\dfrac{1}{3}\cdot\dfrac{3}{5}$

    a. $-\dfrac{1}{5}$ **b.** $-\dfrac{3}{15}$ **c.** $\dfrac{1}{15}$ **d.** $-\dfrac{1}{15}$

10. $\dfrac{4}{3}+\dfrac{x}{4}$

    a. $\dfrac{4x}{7}$ **b.** $\dfrac{16+3x}{12}$ **c.** $\dfrac{4+x}{7}$ **d.** $\dfrac{4+x}{12}$

Chapter 4 Practice Test B (cont'd)

11. $\dfrac{1}{7} - \dfrac{2}{x}$

    a. $\dfrac{-1}{7x}$ b. -1 c. $\dfrac{x-14}{7x}$ d. $\dfrac{x-2}{7x}$

12. $-\dfrac{4}{5}\left(-\dfrac{5}{6}\right)$

    a. $\dfrac{2}{3}$ b. $-\dfrac{2}{3}$ c. $-\dfrac{1}{2}$ d. $\dfrac{1}{6}$

13. $-\dfrac{5}{9} + \dfrac{3}{5}$

    a. $-\dfrac{2}{45}$ b. $-1\dfrac{7}{45}$ c. $-\dfrac{52}{45}$ d. $\dfrac{2}{45}$

14. $\dfrac{4x}{7} \div \dfrac{x}{14}$

    a. 8 b. $\dfrac{4x}{14}$ c. $\dfrac{2x^2}{49}$ d. $\dfrac{1}{8}$

15. $-\dfrac{3}{5} - \dfrac{7}{9}$

    a. $-\dfrac{8}{15}$ b. $-\dfrac{2}{9}$ c. $-\dfrac{10}{45}$ d. $-1\dfrac{17}{45}$

16. $\dfrac{5}{6} - \dfrac{3}{4} + \dfrac{7}{12}$

    a. $\dfrac{9}{22}$ b. $\dfrac{9}{14}$ c. $\dfrac{2}{3}$ d. $\dfrac{3}{4}$

17. $4\dfrac{3}{4} + 2\dfrac{3}{5}$

    a. $6\dfrac{7}{20}$ b. $7\dfrac{7}{20}$ c. $6\dfrac{2}{3}$ d. $7\dfrac{2}{3}$

18. $4 - 2\dfrac{3}{7}$

    a. $1\dfrac{4}{7}$ b. $2\dfrac{4}{7}$ c. $2\dfrac{3}{7}$ d. $1\dfrac{3}{7}$

Chapter 4 Practice Test B (cont'd)

19. $-3\frac{3}{4} \div 1\frac{9}{16}$

    a. $-2\frac{3}{16}$ **b.** $-3\frac{3}{4}$ **c.** $-2\frac{2}{5}$ **d.** $-\frac{5}{12}$

20. $3\frac{3}{8} - 1\frac{3}{4}$

    a. $2\frac{3}{8}$ **b.** $1\frac{5}{8}$ **c.** $1\frac{7}{8}$ **d.** $2\frac{3}{4}$

21. $\frac{1}{3} \div \left(\frac{3}{5} \cdot \frac{1}{4}\right)$

    a. $\frac{1}{20}$ **b.** $\frac{5}{36}$ **c.** $\frac{1}{12}$ **d.** $2\frac{2}{9}$

22. $\left(\frac{1}{2}\right)^2 + \left(-\frac{2}{3}\right)^2$

    a. $\frac{1}{9}$ **b.** $-\frac{7}{36}$ **c.** $\frac{25}{36}$ **d.** $-\frac{1}{9}$

Simplify each complex fraction.

23. $\dfrac{\frac{2}{3}}{\frac{5}{6}}$

    a. $\frac{4}{5}$ **b.** $1\frac{1}{4}$ **c.** $\frac{5}{9}$ **d.** $\frac{2}{3}$

24. $\dfrac{\frac{4}{5} - 1}{\frac{3}{5} + \frac{1}{3}}$

    a. $-\frac{3}{14}$ **b.** $\frac{11}{14}$ **c.** -1 **d.** 1

Solve each equation.

25. $-\frac{4}{5}x = \frac{5}{12}$

    a. $1\frac{13}{60}$ **b.** $-\frac{25}{48}$ **c.** $-\frac{1}{3}$ **d.** $-\frac{23}{60}$

Chapter 4 Practice Test B (cont'd)

26. $x - \dfrac{2}{3} = -\dfrac{1}{3}$

    a. -1 **b.** $\dfrac{1}{3}$ **c.** $-\dfrac{1}{2}$ **d.** $\dfrac{1}{2}$

27. $\dfrac{x}{2} - 1 = \dfrac{1}{3} + \dfrac{1}{4}$

    a. $3\dfrac{1}{6}$ **b.** $\dfrac{6}{19}$ **c.** 1 **d.** $1\dfrac{1}{3}$

Evaluate each expression for the given replacement values.

28. $2x - y$ for $x = -\dfrac{1}{4}$ and $y = \dfrac{1}{3}$

    a. $\dfrac{5}{6}$ **b.** $-\dfrac{5}{6}$ **c.** $\dfrac{1}{24}$ **d.** $\dfrac{5}{24}$

29. $x \div y$ for $x = -\dfrac{1}{5}$ and $y = \dfrac{4}{5}$

    a. 4 **b.** $-\dfrac{4}{25}$ **c.** $-\dfrac{1}{4}$ **d.** -4

30. A carpenter cuts a piece $4\dfrac{2}{3}$ feet off of a board that was 6 feet long. How much is left?

    a. $1\dfrac{1}{3}$ feet   **b.** $2\dfrac{1}{3}$ feet   **c.** $\dfrac{1}{3}$ foot   **d.** $1\dfrac{5}{6}$ feet

Use the rectangle for problems 31 and 32.

$\dfrac{1}{4}$ foot

$\dfrac{2}{3}$ foot

31. Find the perimeter.

    a. $\dfrac{1}{6}$ foot   **b.** $\dfrac{11}{12}$ foot   **c.** $\dfrac{1}{3}$ foot   **d.** $1\dfrac{5}{6}$ feet

32. Find the area.

    a. $\dfrac{1}{6}$ square foot   **b.** $\dfrac{11}{12}$ square foot   **c.** $\dfrac{1}{3}$ square foot   **d.** $1\dfrac{5}{6}$ square feet

33. Ralph wants to cut a 60 inch board into pieces that were $2\dfrac{2}{5}$ inches. How many pieces could he cut?

    a. 22 pieces   **b.** 30 pieces   **c.** 25 pieces   **d.** 14 pieces

5.1 Introduction to Decimals

<table>
<tr><td>

Learning Objectives

  A. Know the meaning of place value for a decimal number and write decimals in words.

  B. Write decimals in standard form.

  C. Write decimals as fractions.

  D. Compare decimals.

  E. Round decimals to a given place value.

</td></tr>
</table>

Objective A Know the meaning of place value for a decimal number and write decimals in words.

    Example 1 Write each decimal in words.

    (a) 0.3   (b) −12.62   (c) 3.157

    Example 2 Write the decimal in the following sentence in words. The Carmen Lucia Ruby in the Smithsonian Museum of Natural History is 23.1 carats.

    Example 3 Write the decimal in the following sentence in words. The Hope Diamond in the Smithsonian Museum of Natural History is 45.52 carats.

    Example 4 Fill in the check to Barnes and Noble for your purchase of $27.85.

                                            Date_____

Pay to _____ $_____

_____ Dollars

_____

1a) three tenths, b) negative twelve and sixty-two hundredths, c) three and one hundred fifty-seven thousandths 2) twenty-three and one tenth, 3) forty-five and fifty-two hundredths

4) Twenty-seven and $\dfrac{85}{100}$

Example 5.1 (cont'd)

Objective B Write decimals in standard form.

Write each decimal in standard form.

Example 5   thirty-seven and thirteen hundredths

Example 6 five and three hundredths

Objective C Write decimals as fractions.

Example 7 Write 0.21 as a fraction.

Example 8 Write 6.3 as a mixed number.

Write each decimal as a fraction in lowest terms.

Example 9 0.375

Example 10 2.36

Example 11 −41.003

Objective D Compare decimals.

Write < or > to form a true statement.

Example 12 0.462   0.465

Example 13 0.016   0.10

Example 14 −0.061   −0.62

Objective D Round decimals.

Example 15 Round 6.60526 to the nearest tenth.

Example 16 Round −0.1365 to the nearest hundredths place.

Example 17 $\pi \approx 3.14159265$ Round π to the nearest thousandth.

Example 18 Round $57,293.91 to the nearest dollar.

5) 37.13, 6) 5.03, 7) $\frac{21}{100}$, 8) $6\frac{3}{10}$, 9) $\frac{3}{8}$, 10) $2\frac{9}{25}$, 11) $-41\frac{3}{1000}$, 12) <, 13) <, 14) >,
15) 6.6, 16) −0.14, 17) 3.142, 18) $57,294

Practice Set 5.1

Objective A Know the meaning of place value for a decimal and write decimals in words.

Write each decimal number in words.

1. 6.48   1. _____

2. 24.375   2. _____

3. −0.17   3 _____

4. 2000.001   4. _____

5. 106.4   5. _____

Objective B Write decimals in standard form.

Write each decimal in standard form.

6. three and four tenths   6. _____

7. five and eight hundredths   7. _____

8. seventy-two and three hundred seventeen thousandths   8. _____

9. negative one and six hundredths   9. _____

10. two and forty-five hundredths   10. _____

Objective C Write decimals as fractions.

Write each decimal as a fraction or mixed number in simplest form.

11. 0.9   11. _____

12. 0.8   12. _____

13. 0.16   13. _____

Practice Set 5.1 (cont'd)

14. 2.35

14. _____

15. 3.004

15. _____

Objective D Compare decimals.

Insert < or > between each pair of numbers to form a true statement.

16. 0.315   0.317

16. _____

17. 0.21   0.168

17. _____

18. −0.045   −0.049

18. _____

Objective E Round decimals to a given place value.

Round each decimal to the given place value.

19. 0.3698, nearest hundredths

19. _____

20. 3.615, nearest tenth

20. _____

21. −2.0481, nearest hundredths

21. _____

22. 5.12987, nearest tenths

22. _____

23. $37.79, nearest dollar

23. _____

Extensions

24. Write a four digit number that will round to 15.2.

24. _____

25. Write these numbers from smallest to largest.
0.001, 0.135, 0.047, 0.136, 0.072

25. _____

Name: Date:

Instructor: Section:

5.2 Adding and Subtracting Decimals

Learning Objectives
- A. Add or subtract decimals.
- B. Estimate when adding or subtracting decimals.
- C. Evaluate expressions with decimal replacement values.
- D. Simplify expressions containing decimals.
- E. Solve problems that involve adding or subtracting decimals.

Objective A Add or subtract decimals.

Example 1 Add: $17.20 + 1.365$

Example 2 Add: $23.61 + 4.792 + 121.394$

Example 3 Add: $17 + 3.12$

Example 4 Add: $4.91 + (-5.94)$

Example 5 Subtract: $7.2 - 0.89$

Example 6 Subtract: $37 - 12.04$

Example 7 Subtract 6 from 12.18.

Example 8 Subtract: $-9.2 - 1.41$

1) 18.565, 2) 149.796, 3) 20.12, 4) −1.03, 5) 6.31, 6) 24.96, 7) 6.18, 8) −10.61

Examples 5.2 (cont'd)

Example 9 Subtract: $-3.21 - (-3.3)$

Objective B Estimate when adding or subtracting decimals.

Example 10 Add or subtract as indicated. Then estimate to see if the answer is reasonable by rounding the given numbers , and adding or subtracting the rounded numbers.

a) $37.5 + 12.8$ b) $42.3 - 37.84$

Objective C Evaluate expressions with decimal replacement values.

Example 11 Evaluate $x - y$ for $x = 1.4$ and $y = 0.95$.

Example 12 Is 0.8 a solution of the equation $1.7 = x + 0.9$?

Objective D Simplify expressions containing decimals.

Example 13 Simplify by combining like terms. $3.2x + 4.91 + 2.91x - 6$

Objective E Solve problems that involve adding or subtracting decimals.

Example 14 Find the total monthly cost of owning and operating an automobile given the expenses shown.

Monthly car payment: $321.91
Monthly insurance cost: $125.19
Average gasoline per month $84.72

Example 15 The Golden Jubilee Diamond is 545.67 carats. The Hope Diamond is 45.52 carats. How much larger is the Golden Jubilee Diamond than the Hope Diamond?

9) 0.09, 10a) 50.3, b) 4.46, 11) 0.45, 12) yes, 13) 6.11x – 1.09, 14) $531.82, 15) 500.15 carats

Practice Set 5.2

Objective A Add or subtract fractions.

Add or subtract.

1. $4.6 + 0.23$

1. _____

2. $5.61 + 1.04$

2. _____

3. $6.1 - 4.29$

3. _____

4. $378.92 + 147.21$

4. _____

5. $-1.72 + (-7.21)$

5. _____

6. $4.21 - 7.3$

6. _____

Objective B Estimate when adding or subtracting decimals.

Add or subtract. Estimate to see if the answer is reasonable.

7. $12.5 - 3.2$

7. _____

8. $100 - 2.3$

8. _____

9. $27.9 + 34.91$

9. _____

10. $7 - 0.00031$

10. _____

11. $-8.43 + 6.1$

11. _____

Practice Set 5.2 (cont'd)

Objective C Evaluate expressions with decimal replacement values.

Evaluate each expression for $x = 1.7$, $y = 7$, and $z = -3.4$.

12. $x + y$ 12. _____

13. $x - z$ 13. _____

14. $x + y + z$ 14. _____

15. Is 2.3 a solution to $x - 4.7 = 1.83$?   15. _____

16. Is -1.3 a solution to $x + 1.7 = 2x - 3$?   16. _____

Objective D Simplify expressions containing decimals.

Simplify by adding like terms.

17. $1.2x - 3.7 + 8.6x - 2.1$ 17. _____

18. $-7.31x - 4.71 - 2.95x - 3.04$ 18. _____

Objective E Solve problems that involve adding or subtracting decimals.

19. Find the perimeter of a square with sides of 1.63 in.   19. _____

20. Kelly's GPA for his first semester at college was   20. _____
2.98. His GPA for his second semester was 3.46.
How much better was his GPA his second semester?

Extensions

21. Find the diameter of the inner circle.   21. _____

‹—— 3.65 inches ——›

1.14
inches

5.3 Multiplying Decimals and Circumference of a Circle

Name: _____ **Date:** _____
Instructor: _____ **Section:** _____

5.3 Multiplying Decimals and Circumference of a Circle

Learning Objectives
  A. Multiply decimals.
  B. Estimate when multiplying decimals.
  C. Multiply decimals by ten.
  D. Evaluate expressions with decimal replacement values.
  E. Find the circumference of a circle.
  F. Solve problems by multiplying decimals.

Vocabulary
circumference

Objective A Multiply decimals.

Example 1 Multiply: 5.62×1.3

Example 2 Multiply: 0.061×1.23

Example 3 Multiply: $(-1.4)(2.3)$

Objective B Estimate when multiply decimals.

Example 4 Multiply: 32.15×1.62. Then estimate to see whether the answer is reasonable.

Objective C Multiply decimals by powers of 10.

Multiply.

Example 5 3.91×10

_1) 7.306, 2) 0.07503, 3) −3.22, 4) 52.083, 5) 39.1_

Martin-Gay **Prealgebra** edition 5

169

Examples 5.3 (cont'd)

Example 6 42.312×100

Example 7 $(-2.61)(1000)$

Example 8 3.6×0.1

Example 9 478×0.01

Example 10 $(-3.2)(0.0001)$

Example 11 In 2005, the population of Texas was 22.9 million. Write this number in standard notation.

Objective D Evaluate expressions with decimal replacement values.

Example 12 Evaluate xy for $x = 1.5$ and $y = 0.32$.

Example 13 Is -3 a solution to $4.1a = -12.3$?

Objective E Find the circumference of a circle.

Example 14 Find the circumference of a circle whose radius is 3 inches. Then use the approximation 3.14 for π to approximate the circumference.

Objective F Solve problems by multiplying decimals.

Example 15 A handyman is hired to paint an apartment with paint costing $8.74 per gallon. If the job requires 4 gallons of paint, what is the total cost of the paint?

6) 4231.2, 7) −2610, 8) 0.36, 9) 4.78, 10) −0.00032, 11) 22,900,000, 12) 0.48, 13) yes, 14) 6π inches, 18.84 inches, 15) $34.96

Practice Set 5.3

Use the choices below to fill in each blank.

    circumference   **left**

      right   **sum**

1. In multiplication of decimals, the number of decimal places in the product is equal to the _____ of the number of decimals in the factors.

2. When multiplying by powers of ten that are larger than 1, (10, 100, 1000 . . .) we move the decimal point to the _____ the number places that there are zeros in the power of 10.

3. When multiplying by powers of ten that are less than 1 (0.1, 0.01, 0.001 . . .), we move the decimal point the to _____ the same number of places are there are decimal places in the power of ten.

4. The _____ is the distance around a circle.

Objective A Multiply decimals.

Multiply.

5. 0.25×0.6

                                                       5. _____

6. 3.6×0.12

                                                       6. _____

7. 4×0.061

                                                       7. _____

8. $(-2.7)(-4.8)$

                                                       8. _____

Objective B Estimate when multiplying.

Multiply. Estimate to see if the answer is reasonable.

9. 0.295×0.6

                                                       9. _____

Practice Set 5.3 (cont'd)

Objective C Multiply by powers of ten.

10. 3.1×100

10. _____

11. 0.006×100

11. _____

12. 39.4×0.1

12. _____

13. 467×0.00001

13. _____

14. The distance from the earth to the sun is 92.96 million miles. Write this number in standard notation.

14. _____

Objective D Evaluate expressions with decimal replacement values.

15. Evaluate for $x = 1.6$ and $y = -2.3$. xy

15. _____

16. Is 0.6 a solution to $3.6x = 2$?

16. _____

Objective E Find the circumference of a circle.

Find the circumference of the circle. Then use the approximation 3.14 for π and approximate each circumference.

17. Circle with diameter of 24 in.

17. _____

_____

Objective F Solve problems by multiplying decimals.

18. A slice of bread has 72.3 calories. How many calories are in 22 slices of bread?

18. _____

19. A slice of bread has 1.2 grams of fat. How many grams of fat are in 4 slices of bread?

19. _____

Extensions

20. Find the circumference of the inside circle.

20. _____

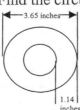

3.65 inches

1.14 inches

5.4 Dividing Decimals

Learning Objectives
    A. Divide decimals.
    B. Estimate when dividing decimals.
    C. Divide decimals by powers of 10.
    D. Evaluate expressions with decimal replacement values.
    E. Solve problems by dividing decimals.

Objective A Divide decimals.

    Divide

Example 1 $5.76 \div 3$

Example 2 $21\overline{)7.35}$

Example 3 $4.41 \div 105$

1) 1.92, 2) 0.35, 3) 0.042

Examples 5.4 (cont'd)

Example 4 5.457 ÷ 1.7

Example 5 0.0756 ÷ 0.021

Example 6 19.3 ÷ 0.47 Round your answer to the nearest hundredth.

4) 3.21, 5) 3.6, 6) 41.06

Name: Date:
Instructor: Section:

Examples 5.4 (cont'd)

Objective B Estimate when dividing decimals.

Example 7 Divide. Then estimate to see whether the result is reasonable.
$200.22 \div 21.3$

Objective C Divide decimals by powers of 10.

Divide.

Example 8 $3.61 \div 10$

Example 9 $4.216 \div 100$

Objective D Evaluate expressions with decimal replacement values.

Example 10 Evaluate $x \div y$ for $x = 3.6$ and $y = 0.004$.

7) 9.4, 8) 0.361, 9) 0.04216, 10) 900

Examples 5.4 (cont'd)

Example 11 Is 46 a solution to $\dfrac{y}{100} = 4.6$?

Objective E Solve problems by dividing decimals.

Example 12 A gallon of paint covers 250-square-foot area. If you wish to paint an apartment whose walls total to 1625 square feet of area, how much paint do you need? Give answer in whole gallons.

11) no, 12) 7 gallons

Practice Set 5.4

Use the choices below to fill in each blank.

dividend   **divisor**
left   **right**

1. To divide a decimal number by a power of 10 that is greater than 1, we move the decimal point in the number to the _____ the same number of places as there are zeros in the power of 10.

2. To divide by 0.36, move the decimal in the _____ and the _____ two places to the _____.

Objective A Divide decimals.

Divide.

3. $2\overline{)4.78}$ 3. _____

4. $6\overline{)0.39}$ 4. _____

5. $0.24\overline{)0.768}$ 5. _____

6. $0.12\overline{)0.414}$ 6. _____

Practice Set 5.4 (cont'd)

7. $45 \div 0.0009$ 7. _____

8. $4.307 \div 0.73$ 8. _____

9. $4.2 \div 0.06$ 9. _____

Objective B Estimate when dividing decimals.

Divide. Then estimate to check the reasonableness of the answer.

10. $3.6\overline{)30.24}$ 10. _____

Name: _____   Date: _____
Instructor: _____   Section: _____

Practice Set 5.4 (cont'd)

11. $3.18\overline{)8.268}$

12. _____

12. Round to the nearest tenth. $0.614 \div 0.17$

12. _____

Objective C Divide decimals by powers of ten.

13. $\dfrac{3.621}{100}$

13. _____

14. $\dfrac{41.59}{10}$

14. _____

15. $-13.5 \div (-1000)$

15. _____

Martin-Gay **Prealgebra** edition 5

179

Practice Set 5.4 (cont'd)

Objective D Evaluate expressions with decimal replacement values.

16. Evaluate $x \div y$ for $x = 0.288$ and $y = 0.8$.   16. _____

17. Evaluate $\dfrac{x}{y}$ for $x = 1.6$ and $y = 0.8$.   17. _____

Objective E Solve problems by dividing decimals.

18. Rebecca is painting the walls of a house that has a   18. _____
total wall area of 2462 square feet. A gallon of paint
will cover 210 square feet. How many gallons of
paint must she buy. Round the answer to whole gallons.

19. There are approximately 39.37 inches in 1 meter.   19. _____
How many meters, to the nearest tenth of a meter,
are there in 500 inches?

Extensions

20. The radius of a circular irrigation system is 1250 yards.  **20.** _____
What is the circumference of the area that can be
watered? Approximate using $\pi \approx 3.14$.

5.5 Fractions, Decimals, and Order of Operations

Learning Objectives
 A. Write fractions as decimals.
 B. Compare fractions and decimals.
 C. Simplify expressions containing decimals and fractions using order of operations.
 D. Solve problems containing fractions and decimals.
 E. Evaluate expressions given decimal replacement values.

Objective A Write fractions as decimals.

Example 1 Write $\dfrac{3}{4}$ as a decimal.

Example 2 Write $-\dfrac{3}{8}$ as a decimal.

Example 3 Write $\dfrac{1}{6}$ as a decimal.

Example 4 Write $\dfrac{17}{3}$ as a decimal. Round to the nearest hundredth.

1) 0.75, 2) –0.375, 3) $0.1\overline{6}$, 4) 5.67

Practice Set 5.5 (cont'd)

Example 5 Write $1\frac{2}{5}$ as a decimal.

Example 6 Write $\frac{1}{2}$ as a decimal.

Example 7 Write $\frac{3}{25}$ as a decimal.

Objective B Compare fractions and decimals.

Insert <, >, or = to form a true statement.

Example 8 $\frac{2}{5}$ 0.42

5) 1.4, 6) 0.5, 7) 0.12, 8) <

Practice Set 5.5 (cont'd)

Example 9 $0.\overline{5}$ $\dfrac{5}{9}$

Example 10 Write the numbers in order from smallest to largest. $\dfrac{3}{5}$, 0.688, $\dfrac{2}{3}$

Objective C Simplify expressions containing decimals and fractions using order of operations.

Simplify
Example 11 $73.46 \div 100 \times 10$

Example 12 $-0.4(4.6 - 0.3)$

9) =, 10) $\dfrac{3}{5}$, $\dfrac{2}{3}$, *0.688*, 11) *7.346*, 12) *−1.72*

Practice Set 5.5 (cont'd)

Example 13 $(1.2)^2 - 4.6$

Example 14 $\dfrac{3.69 - (0.4)^2 \times 10}{0.2}$

Example 15 The area of a triangle is Area $= \dfrac{1}{2} \cdot$ base $\cdot$ height. Find the area of the triangle shown.

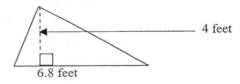

6.8 feet

4 feet

Objective E Evaluate expressions given decimal replacement values.

Example 16 Evaluate $3x - 6$ for $x = 4.1$.

13) −3.16, 14) 10.45, 15) 13.6 square feet, 16) 6.3

Practice Set 5.5

Objective A Write fractions as a decimals.

Write each number as a decimal.

1. $\dfrac{12}{25}$ 1. _____

2. $\dfrac{9}{20}$ 2. _____

3. $\dfrac{5}{12}$ 3. _____

4. $\dfrac{7}{11}$ 4. _____

5. $3\dfrac{3}{5}$ 5. _____

6. $\dfrac{7}{25}$ 6. _____

Practice Set 5.5 (cont'd)

7. $4\dfrac{3}{7}$ to the nearest hundredth.   7. _____

Objective B Compare fractions and decimals.

Insert <, >, or = to form a true statement.

8. 0.63 $\dfrac{48}{75}$ 8. _____

9. 0.667 $\dfrac{2}{3}$ 9. _____

10. $\dfrac{17}{23}$ 0.739   10. _____

11. Write the numbers in order from smallest to largest.   11. _____
$$0.411, \ 0.43, \ \frac{11}{25}$$

Practice Set 5.5 (cont'd)

Objective C Simplify expressions containing decimals and fractions using order of operations.

Simplify each expression.

12. $(0.4)^2 + 1.2$

12. _____

13. $(-3.6)(2) - 4.8$

13. _____

14. $0.6 - 0.4(3.6 - 4.2)$

14. _____

Objective D Solve area problems containing fractions and decimals.

Find the area of each figure.

15.

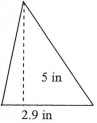

5 in

2.9 in

15. _____

16.

3/5 yards

1.6 yards

16. _____

Practice Set 5.5 (cont'd)

Objective E Evaluate expressions given decimal replacement values.

17. Evaluate $x - y^2$ for $x = 1.6$ and $y = 3.1$ **17.** _____

18. Evaluate $2x - y$ for $x = -1.6$ and $y = 4.2$.   **18.** _____

19. Evaluate $\dfrac{x}{y} + 2x$ for $x = -3.6$ and $y = 1.2$.   **19.** _____

Extensions
Use the bar graph to answer the following questions.

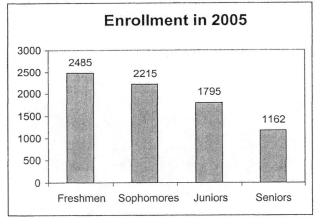

20. What fraction of the students are freshmen?   **20.** _____
Round answers to the nearest tenth.

21. What fraction of the students are seniors?   **21.** _____
Round answers to the nearest tenth.

5.6 Equations Containing Decimals

Learning Objectives
   A. Solve equations containing decimals.

Objective A Solve equations containing decimals.

Solve.

Example 1 $x - 2.4 = 1.6$

Example 2 $-3y = -1.41$

Example 3 $2.3x - 1.4 = 5.5$

Example 4 $3x - 4.6 = 5x + 4.2$

Example 5 $4(x - 0.28) = -3x + 0.35$

Example 6 $0.3x - 4.6 = 1.7$

1) 4, 2) 0.47, 3) 3, 4) –4.4, 5) 0.21, 6) 21

Practice Set 5.6

Objective A Solve equations containing decimals.

Solve each equation.

1. $x + 3.6 = -4.5$

1. _____

2. $y - 0.4 = 3$

2. _____

3. $2.3a = 8.05$

3. _____

4. $-0.4x = 46$

4. _____

5. $3.7 = x + 12.5$

5. _____

6. $2x - 3.5 = 4.8$

6. _____

Practice Set 5.6 (cont'd)

7. $5x - 2.38 = 3x - 4.92$

7. _____

8. $3(x - 4.2) = 2x + 7$

8. _____

9. $4(x + 1.6) = 16$

9. _____

10. $0.5x + 0.5 = 6$

10. _____

11. $10x - 3.4 = -7x$

11. _____

12. $1.8x - 5 - 1.4x = 9$

12. _____

Practice Set 5.6 (cont'd)

13. $7x - 3.5 = 6x + 4.28$

13. _____

14. $-0.08x = 0.44$

14. _____

15. $7.2x = -9.727$

15. _____

16. $-40x + 2 = 30x - 0.8$

16. _____

17. $0.8x - 12.4 = x - 3.1$

17. _____

18. $y - 7 = 0.9x + 7.4$

18. _____

19. $3(2x - 0.7) = 5x - 3.61$

19. _____

Extensions

20. Write an equation with −2.4 as the solution.

20. _____

5.7 Decimal Applications: Mean, Median, and Mode

Learning Objectives
  A. Find the mean of a list of numbers.
  B. Find the median of a list of numbers.
  C. Find the mode of a list of numbers.

Vocabulary
mean, median, mode

Objective A Find the mean of a list of numbers.

Find the mean.

Example 1 Five students in a driver's education class were given a test of their reactions times. The results are given below. Find the mean.

Student	Terra	Toni	Don	Jason	Michelle
Reaction Time (Seconds)	2.77	1.89	2.25	3.41	1.23

Example 2 Calculate the grade point average. Round to the nearest hundredth.

Course	Grade	Point Value of Grade	Credit Hours
College Algebra	C	2	3
Chemistry	B	3	4
English	A	4	3
PE	A	4	1
History	C	2	3

Objective B Find the median of a list of numbers.

Example 3 Find the median of the following list of numbers.
          1.6, 2.3, 3.4, 1.8, 2.9, 3.6, 4.1

1) 2.31, 2) 2.86, 3) 2.9

Example 5.7 (cont'd)

Example 4 Find the median of the following list of scores.
86, 68, 85, 94, 96, 84

Objective C Find the mode of a list of numbers.

Example 5 Find the mode of the list of numbers.
9, 3, 6, 8, 9, 5, 11, 17

Example 6 Find the median and the mode of the following list of numbers.
112, 136, 142, 198, 178, 189, 142

4) 85.5, 5) 9, 6) median = 142, mode = 142

Practice Set 5.7

Use the choices below to fill in each blank.

    average   **mean**   **median**   **mode**

1. The number that appears the most often in a list of numbers is called_____.

2. Another word for "mean" is _____.

3. The _____ is the middle value of a set of numbers.

4. Is the mean, median, or mode used to decide your grade at the end the semester?
   _____

Objective A Find the mean of a list of numbers..

Find the mean.

5. 17, 16, 19, 27, 27, 30, 32   5. _____

6. 3.7, 4.2, 4.8, 5.2, 5.2, 6.3   6. _____

7. 473, 495, 506, 472, 603, 480, 570, 522   7. _____

8. Round to the nearest tenth.   8. _____
   111, 109, 121, 147, 131, 132, 108, 146, 127

Practice Set 5.7 (cont'd)

9. Find the grade point average to the nearest hundredth.   9. _____

Course	Grade	Point Value of Grade	Credit Hours
College Mathematics	B	3	3
Biology	A	4	4
English	C	2	3
Government	C	2	3

10. Find the grade point average to the nearest hundredth.   10. _____

Course	Grade	Point Value of Grade	Credit Hours
History	D	1	3
Geology	F	0	4
English	C	2	3
PE	A	4	1
History	B	3	3

Objective B Find the median of a list of numbers.

11. 17, 16, 19, 27, 27, 30, 32   11. _____

12. 3.7, 4.2, 4.8, 5.2, 5.2, 6.3   12. _____

13. 473, 495, 506, 472, 603, 480, 570, 523   13. _____

14. 111, 109, 121, 147, 131, 132, 108, 146, 127   14. _____

Practice Set 5.7 (cont'd)

Objective C Find the mode of a list of numbers.

Find the mode.

15. 17, 16, 19, 27, 27, 30, 32   15. _____

16. 3.7, 4.2, 4.8, 5.2, 5.2, 6.3   16. _____

17. 473, 495, 506, 472, 603, 480, 570, 523   17. _____

18. 111, 109, 121, 147, 131, 132, 108, 146, 127   18. _____

Extensions

Find the missing numbers in each set of numbers.

19. 116, 117, _____, _____, 121, The mode is 121. The median is 118

20. _____, _____, _____, 25, _____, The mean is 23. The median is 23. The mode is 20.

Chapter 5 Vocabulary Reference Sheet

Term	Definition	Example
	Section 5.1	
Decimal	A decimal represents a part of a whole.	0.8 is eight tenths.
	Section 5.3	
Circumference	The distance around a circle.	The circumference of a is $C = \pi d$ or $C = 2\pi r$.
	Section 5.7	
Mean	The sum of a list items divided by the number of items.	For the numbers 3, 5, 6, 7, 9, 9, 10, the average is 7.
Median	The middle value of a list of numbers.	For the numbers 3, 5, 6, 7, 9, 9, 10, the median is 7. For the numbers 16, 17, 18, 18, 20, 22, 24, 25, the average of 18 and 20 (19) is the median.
Mode	The value that appears the most often in a list of numbers.	For the numbers 3, 5, 6, 7, 9, 9, 10, the mode is 9. For the numbers 70, 75, 82, 86, 94, there is no mode.

Chapter 5 Practice Test A

Write each decimal as indicated.

1. 36.195 in words.   1. _____

_____

2. one and sixty-five thousandths in standard form.   2. _____

Perform each indicated operation.

3. $3.96 + 3.06 + 1.195$ 3. _____

4. $-7.98 + 3.62$ 4. _____

5. $-5.62 - 8.113$ 5. _____

6. 2.1×4.13 6. _____

7. $3.927 \div (-.17)$ 7. _____

Round each decimal to the indicated place value.

8. 47.29368 to the nearest tenth   8. _____

9. 0.9249 to the nearest hundredth   9. _____

Insert $<$, $>$, or $=$ between each pair of numbers to form a true statement.

10. 36.012   36.015   10. _____

11. $\dfrac{3}{7}$ 0.428   11. _____

Write each decimal as a fraction or as a mixed number.

12. 0.76   12. _____

13. -5.24 13. _____

Chapter 5 Practice Test A (cont'd)

Write each fraction as a decimal.

14. $-\dfrac{39}{75}$

14. _____

15. $\dfrac{9}{13}$, nearest hundredth

15. _____

Simplify.

16. $(1.3)^2 - 4.9$

16. _____

17. $\dfrac{0.16 + 0.23}{-0.03}$

17. _____

18. $4.6x - 1.92 - 1.7x + 9.7$

18. _____

Solve.

19. $1.7x - 1.2 = 1.2x$

19. _____

20. $3(x - 0.2) = 2x + 0.48$

20. _____

Chapter 5 Practice Test A (cont'd)

Find the mean, median, and mode for each list of numbers.

21. 17, 29, 47, 53, 58

21. mean_____

median_____

mode_____

22. 9, 10, 12, 36, 48, 48, 58, 59

22. mean_____

median_____

mode_____

Find the grade point average to the nearest hundredth.

23.

23. _____

Course	Grade	Point Value of Grade	Credit Hours
College Algebra	B	3	3
Anatomy	B	3	4
English	A	4	3
PE	C	2	1
History	C	2	3

24. Find the area.

24. _____

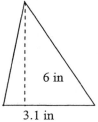

6 in

3.1 in

25. Find the circumference of a circle with radius of 7 yards. Then use 3.14 for an approximation of π to find the approximate circumference.

25. _____

Chapter 5 Practice Test B

Write each decimal as indicated.

1. 23.039 in words
   a. twenty-three and thirty-nine hundredths
   b. twenty-three and thirty-nine thousandths
   c. twenty-three and thirty-nine ten thousandths
   d. twenty-three and thirty-nine tenths

2. three and forty-nine hundredths in standard form.
   a. 3.049   b. 49.3   c. 3.49   d. 3.0049

Perform each indicated operation.

3. $2.36 + 0.495 + 1.6$
   a. 4.45   b. 6.37   c. 4.455   d. 7.47

4. $8.62 - 9.41$
   a. -0.79 b. 0.79   c. 1.21   d. -1.21

5. $-3.75 - 0.431$
   a. -3.281 b. -4.181 c. 3.319   d. -3.319

6. 1.6×4.9
   a. 6.74   b. 67.4   c. 7.84   d. 78.4

7. $0.714 \div (-0.21)$
   a. 0.034   b. 0.34   c. 3.4   d. -3.4

Round each decimal to the indicated place value.

8. 33.4978 to the nearest tenth
   a. 33.4   b. 33.49   c. 33.5   d. 33.498

9. 0.3692 to the nearest hundredth
   a. 0.36   b. 0.369   c. 0.4   d. 0.37

Insert <, >, or = to form a true statement.

10. 4.621   4.609
    a. <   b. >   c. =

Chapter 5 Practice Test B (cont'd)

11. $\dfrac{3}{7}$ 0.428

     a. $<$ **b.** $>$ **c.** $=$

Write each decimal as a fraction or mixed number in simplest form.

12. 0.45

     a. $4\dfrac{5}{10}$ **b.** $4\dfrac{1}{2}$ **c.** $\dfrac{9}{20}$ **d.** $\dfrac{45}{100}$

13. -3.46

     a. $-3\dfrac{46}{100}$ **b.** $-34\dfrac{6}{10}$ **c.** $-34\dfrac{3}{5}$ **d.** $-3\dfrac{23}{50}$

Write each fraction as a decimal.

14. $\dfrac{8}{25}$

     a. 0.32   **b.** 0.3   **c.** 0.315   **d.** 0.426

15. $\dfrac{4}{57}$ to the nearest hundredth

     a. 0.714   **b.** 0.7   **c.** 0.72   **d.** 0.07

Simplify.

16. $(2.1)^2 - 3.8$

     a. 2.89   **b.** 0.4   **c.** 0.61   **d.** -0.61

17. $\dfrac{0.25 + 0.1}{-0.7}$

     a. 5   **b.** -5 **c.** -0.05 **d.** -0.5

18. $3.6x + 1.42 - 1.81x - 0.34$

     a. $-3.95x + 3.4$ **b.** $2.87x$

     c. $-1.79x + 1.08$ **d.** $-1.79x + 1.08$

Chapter 5 Practice Test B (cont'd)

Solve.

19. $7.2x - 3.6 = 7.5x$
    a. −12   **b.** 12   **c.** 1.2   **d.** −1.2

20. $5(x + 0.05) = 3x - 0.97$
    a. 0.61   **b.** −0.61   **c.** −0.46   **d.** 0.46

Use the following list of numbers for problems 21 - 23.
13, 25, 28, 32, 41, 41, 51

21. Find the mean.
    a. 41   **b.** 32   **c.** 33   **d.** 139

22. Find the median.
    a. 41   **b.** 32   **c.** 33   **d.** 139

23 Find the mode.
    a. 41   **b.** 32   **c.** 33   **d.** 139

24. Find the area.

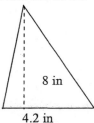

8 in

4.2 in

    a. 33.6 square in   **b.** 16.8 square in   **c.** 4.2 square in   **d.** 12. square in

25. Find the circumference. Then use 3.14 for an approximation of π to find the
    approximate circumference.

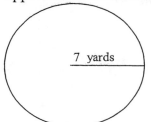

7 yards

    a. 18.84 yards   **b.** 21.98 yards   **c.** 43.96 yards   **d.** 10.14 yards

6.1 Ratios and Rates

Learning Objectives
   A. Write ratios as fractions.
   B. Write rates as fractions.
   C. Find unit rates.
   D. Find unit prices.

Vocabulary
ratio, unit rate, unit price

Objective A Write ratios as fractions.

Example 1 Write the ratio 17 to 32 using fractional notation.

Example 2 Write the ratio $24 to $18 as a fraction in simplest form.

Example 3 Write the ratio of 3.6 to 4.16 as a fraction in simplest form.

Example 4 Write the ratio of $2\frac{1}{10}$ to $\frac{3}{5}$ as a fraction in simplest form.

1) $\frac{17}{32}$, 2) $\frac{4}{3}$, 3) $\frac{45}{52}$, 4) $\frac{7}{2}$

Example 6.1 (cont'd)

Example 5 From the circle graph of a family's monthly expenses, write the ratio of housing to transportation as a fraction in simplest form.

Monthly Expenses

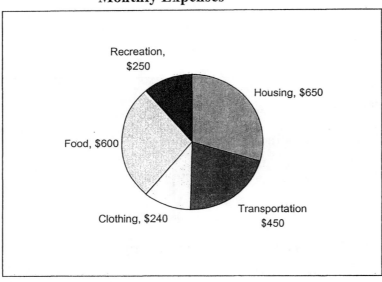

Example 6 Given the rectangle shown:
   a) Find the ratio of its width (shorter side) to its length (longer side).
   b) Find the ratio of its length to its perimeter.

8 feet

5 feet

5) $\frac{13}{9}$, 6a) $\frac{5}{8}$, b) $\frac{4}{13}$

Examples 6.1 (cont'd)

Objective B Write rates as fractions.

Write each rate as a fraction in simplest form.

Example 7 20 miles in 30 minutes

Example 8 430 miles on 15 gallons

Objective C Find unit rates.

Example 9 Write as a unit rate: $12,740 every 5 months

Example 10 Write as a unit rate: 336 miles every 12 gallons

Objective D Find unit price.

Example 11 A store charges $5.88 for 6 ounces of nuts. What is the unit price in dollars per ounce?

Example 12 Approximate each unit price to decide which is the better buy: 6 cans of soda for $1.47 or 12 cans for $3.12.

7) $\dfrac{2 \ miles}{3 \ minutes}$, 8) $\dfrac{86 \ miles}{3 \ gallons}$, 9) *$2548 per month*, 10) *28 miles per gallon*, 11) *$.98 per ounce*,
12) *6 for $1.47*

Practice Set 6.1

Use the choices below to fill in each blank.

       division   **ratio**   **unit**

1. A rate with a denominator of 1 is called a _____ rate.

2. The word "per" tells us to use _____.

3. The quotient of two quantities is called a _____.

Objective A Write ratios as fractions.

Write each ratio as a fraction in simplest form.

4. 18 to 28   4. _____

5. 75 to 225   5. _____

6. 6.6 to 1.04   6. _____

7. $3\dfrac{2}{3}$ to $1\dfrac{2}{6}$ 7. _____

8. In a college algebra class there are 18 girls   8. _____
   and 15 boys. Find the ratio of boys to girls.

Practice Set 6.1 (cont'd)

9. In a college algebra class there are 18 girls
   and 15 boys. Find the ratio of girls to the
   total number of students.

9. _____

Objective B Write ratios as fractions.

Write each rate as a fraction in simplest terms.

10. 3 printers for 21 computers

10. _____

11. 9 tables for 36 diners

11. _____

Objective C Find unit rates.

Write each rate as a unit rate.

12. 252 calories in 21 servings

12. _____

13. 20,000 books for 5000 students

13. _____

Use the following information for problems 14 - 16. Jan can read 30 pages in 15
minutes. Pete can read 12 pages in 5 minutes.

14. Find the unit rate for Jan.

14. _____

15. Find the unit rate for Pete.

15. _____

16. Who can read faster?

16. _____

Practice Set 6.1 (cont'd)

Objective D Find unit prices.

17. Find the unit price. 12 bananas for $2.40

17. _____

18. Find the unit price for each to decide with is
    the better buy: Ketchup $3.60 for 20 ounces
                              $2.52 for 12 ounces

18. _____

19. Find the unit price for each to decide with is
    the better buy: Pens $2.40 for 12 pens
                          $1.68 for 8 pens

19. _____

Extensions

20. Fill in the table to calculate miles per gallon to the nearest tenth.

Beginning Odometer Reading	Ending Odometer Reading	Miles Driven	Gallons of Gas Used	Miles per Gallon
12,662	12,846		6.1	
12,846	13,158		10.2	

6.2 Proportions

Learning Objectives
  A. Write sentences as proportions.
  B. Determine whether proportions are true.
  C. Find an unknown number in a proportion.

Vocabulary
proportion, cross product

Objective A Write sentences as proportions.

  Example 1 Write each sentence as a proportion.
    a) 12 goals is to 15 attempts as 8 goals is to 10 attempts.

    b) 21 customers is to 3 tellers as 35 customers is to 5 tellers.

Objective B Determine whether proportions are true.

  Example 2 Is $\dfrac{5}{6} = \dfrac{25}{30}$ a true proportion?

  Example 3 Is $\dfrac{2.1}{7} = \dfrac{1.7}{6}$ a true proportion?

  Example 4 Is $\dfrac{3\frac{1}{6}}{4\frac{3}{4}} = \dfrac{\frac{4}{5}}{1\frac{1}{5}}$ a true proportion?

1a) $\dfrac{12}{15} = \dfrac{8}{10}$, *b)* $\dfrac{21}{3} = \dfrac{35}{5}$, *2) true, 3) false, 4) true*

Name: Date:
Instructor: Section:

Examples 6.2 (cont'd)

Objective C Find an unknown number in a proportion.

   Example 5 Solve $\dfrac{3}{5} = \dfrac{x}{20}$ for x.

   Example 6 Solve $\dfrac{12}{15} = \dfrac{4}{x}$ for x.

   Example 7 Solve for x. $\dfrac{\frac{1}{3}}{\frac{2}{5}} = \dfrac{x}{1\frac{3}{5}}$

   Example 8 Solve for n. $\dfrac{n}{3} = \dfrac{2.5}{1.5}$

   Example 9 Solve for y. $\dfrac{7}{y} = \dfrac{14}{11}$

   Example 10 Solve of x. Round to the nearest hundredth. $\dfrac{1.7}{1.2} = \dfrac{x}{0.4}$

5) 12, 6) 5, 7) $\dfrac{4}{3}$, 8) 5, 9) $\dfrac{11}{2}$, 10) 0.57

Practice Set 6.2

Use the choices below to fill in each blank.

cross product   **proportion**   **ratio**

1. $\dfrac{3}{5} = \dfrac{6}{10}$ is called a _____.

2. $\dfrac{3}{5}$ is called a _____.

3. To solve the proportion $\dfrac{w}{x} = \dfrac{y}{z}$, $wz = xy$ is called the _____.

Objective A Write statements as proportions.

Write each sentence as a proportion.

4. 7 errors is to 25 problems as 35 errors is to 125 problems

4. _____

5. $3\dfrac{1}{2}$ cups of milk is to 7 cups of flour as $7\dfrac{1}{2}$ cups of milk is to 15 cups of flour.

5. _____

6. 6 slices of bread is to 3 sandwiches as 14 slices of bread is to 7 sandwiches.

6. _____

Objective B Determine whether proportions are true.

Determine whether each proportion is true or false.

7. $\dfrac{0.4}{0.5} = \dfrac{0.8}{0.9}$

7. _____

8. $\dfrac{7}{21} = \dfrac{5}{15}$

8. _____

9. $\dfrac{24}{9} = \dfrac{40}{16}$

9. _____

10. $\dfrac{3.1}{3.6} = \dfrac{6.6}{7.2}$

10. _____

Name:

Instructor:

Date:

Section:

Practice Set 6.2 (cont'd)

11. $\dfrac{1\frac{3}{5}}{2\frac{2}{3}} = \dfrac{\frac{6}{7}}{1\frac{1}{4}}$

11. _____

Objective C Find an unknown number in a proportion.

Solve each proportion for the given variable.

12. $\dfrac{x}{4} = \dfrac{9}{12}$

12. _____

13. $\dfrac{-12}{15} = \dfrac{16}{n}$

13. _____

14. $\dfrac{7}{\frac{1}{2}} = \dfrac{28}{x}$

14. _____

15. $\dfrac{-0.3}{0.5} = \dfrac{x}{1.5}$

15. _____

16. $\dfrac{15}{x} = \dfrac{4}{\frac{3}{5}}$

16. _____

17. Solve and round to the nearest tenth.
$\dfrac{3.5}{0.9} = \dfrac{x}{1.2}$

17. _____

Extensions
18. Write a true proportion.

18. _____

19. Use the numbers in the proportion to write two other true proportions.
$\dfrac{3}{4} = \dfrac{6}{8}$

19. _____

20. Solve. $\dfrac{x}{1.5} = \dfrac{0}{2.7}$

20. _____

Martin-Gay **Prealgebra** edition 5

214

6.3 Proportions and Problem Solving

Learning Objectives
   A. Solve problems by writing proportions.

Objective A Solve problems by writing proportion..

Example 1 On a map, 8.5 corresponds to 3 inches. How many miles corresponds to 9 inches?

Example 2 The standard dose of a drug is 8 ml (milliliters) for each 50 pounds of body weight. Find the dose for a 225-pound male.

Example 3 A 20-pound bag of grass seed covers 1500 square feet. How many pounds are needed to cover 30,000 square feet?

1) 25.5 miles, 2) 36 ml, 3) 400 pounds

Practice Set 6.3

Objective A Solve problems by writing proportions.

Write a proportion and solve.

1. A basketball player makes 4 goals for every 5 attempts at the free throw line. If he made 8 goals in one game, how many attempts did he have?

1. _____

2. A basketball player makes 4 goals for every 5 attempts at the free throw line. For the season he had 80 attempts, how many goals did he make?

2. _____

3. A medical school accepts 1 out of every 8 students that apply. If there are 320 applicants, how many will be accepted?

3. _____

4. A medical school accepts 1 out of every 8 students that apply. If 30 are accepted, how many applied?

4. _____

5. On a map, 1.5 inches represents 40 miles. How many miles does 6 inches represent?

5. _____

6. On a map, 1.5 inches represents 40 miles. How many inches would it take to represent 120 miles?

6. _____

Practice 6.3 (cont'd)

7. A Toyota Prius gets 244 miles on 5 gallons. How   7. _____
    many miles would 12 gallons go?

8. A Toyota Prius gets 244 miles in 5 gallons. How   8. _____
    gallons would it take to go 1952 miles?

9. A bag of fertilizer covers 2800 square feet. How many   9. _____
    square feet would 6 bags cover?

10. A bag of fertilizer covers 2800 square feet. How   10. _____
    bags would it take to cover a square that is 140 feet
    on each side?

11. A baseball player gets 4 hits for every 7 times at bat.   11. _____
    How many hits would he get for 49 times at bat?

12. A baseball player gets 4 hits for every 7 times at bat.   12. _____
    If he had 20 hits, how many times at bat did he have?

13. A brownie recipe calls for $\frac{2}{3}$ cup of cocoa and 4   13. _____

    cups of flour to make 30 brownies. How much
    cocoa would it take to make 45 brownies?

Practice Set 6.3 (cont'd)

14. A brownie recipe calls for $\frac{2}{3}$ cup of cocoa and 4 cups of flour to make 30 brownies. If you had 3 cups of cocoa, how many brownies could be made?

14. _____

15. A brownie recipe calls for $\frac{2}{3}$ cup of cocoa and 4 cups of flour to make 30 brownies. If you had 3 cups of cocoa, how much flour would be needed?

15. _____

16. An office uses 20 pens in 4 weeks. How long 100 pens last?

16. _____

17. An office use 20 pens in 4 weeks. How many pens would be needed for one year (52 weeks)?

17. _____

18. Thereis 290 mg of sodium in 21 cheetos. How many cheetos would it take to have 870 mg of sodium?

18. _____

19. There is 290 mg of sodium in 21 cheetos. How many mg of sodium would be in 63 cheetos?

18. _____

Extensions

20. It takes 2.5 cups of milk to make 10 muffins. How cups of milk would it take to make 2 dozen muffins?

20. _____

6.4 Square Roots and the Pythagorean Theorem

<table>
<tr><td>

Learning Objectives
   A. Find the square root of a number.
   B. Approximate square roots.
   C. Use the Pythagorean Theorem.

Vocabulary
perfect square, square roots, radical, hypotenuse, leg

</td></tr>
</table>

Objective A Find the square root of a number.

Find each square root.

Example 1 $\sqrt{36}$

Example 2 $\sqrt{100}$

Example 3 $\sqrt{25}$

Example 4 $\sqrt{1}$

Example 5 $\sqrt{\dfrac{1}{81}}$

Example 6 $\sqrt{\dfrac{16}{25}}$

Objective B Approximate square roots.

Use Appendix A.4 in the text book or a calculator to approximate each square root to the nearest thousandth.

Example 7
a) $\sqrt{37}$ **b)** $\sqrt{90}$

1) 6, 2) 10, 3) 5, 4) 1, 5) $\dfrac{1}{9}$, 6) $\dfrac{4}{5}$, 7a) 6.083, b) 9.487

Examples 6.4 (cont'd)

Objective C Use the Pythagorean Theorem.

Example 8 Find the length of the hypotenuse of the given right triangle.

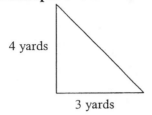

4 yards

3 yards

Example 9 Find the length of the hypotenuse of the given right triangle. Round the length to the nearest whole unit.

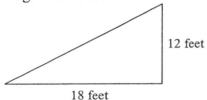

12 feet

18 feet

Example 10 Find the length of the leg in the given right triangle. Give the exact length and a two-decimal place approximation.

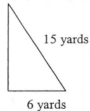

15 yards

6 yards

Example 11 Find the diagonal of the given square. Round answer to the nearest tenth.

200 yards

8) 5 yards, 9) 22 feet, 10) $\sqrt{189} \approx 13.75$ feet, 11) 282.8 yards

Practice Set 6.4

Use the choices below to fill in each blank.

leg   **hypotenuse**   **radical**
perfect square   **square root**

1. The reverse operation of squaring is called taking the _____.

2. The positive square root of a nonnegative number is denoted by the
   _____ sign.

3. The numbers 4, 25, and 64 and called _____.

4. Label the parts of the right triangle.

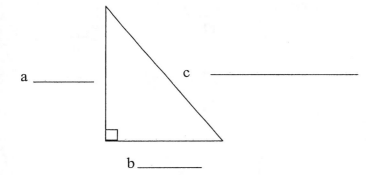

a _____   c   _____

b _____

5. Use the given triangle to fill in the correct letters.
   $x^2 +$ _____ $=$ _____

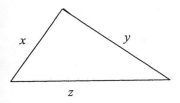

Objective A Find the square root of a number.

Find each square root.

6. $\sqrt{49}$ 6. _____

7. $\sqrt{25}$ 7. _____

Practice Set 6.4 (cont'd)

8. $\sqrt{\dfrac{1}{16}}$

8. _____

9. $\sqrt{\dfrac{36}{49}}$

9. _____

10. $\sqrt{\dfrac{4}{16}}$

10. _____

Concept B Approximate Square Roots
Approximate to the nearest thousandth.

11. $\sqrt{15}$

11. _____

12. $\sqrt{93}$

12. _____

13. $\sqrt{24}$

13. _____

14. $\sqrt{74}$

14. _____

15. $\sqrt{124}$

15. _____

Concept C Use the Pythagorean Theorem

Find the unknown length in each right triangle.

16.

16. _____

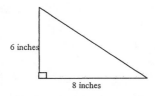

6 inches

8 inches

17. leg = 12, leg = 5

17. _____

18. leg = 7, hypotenuse = 12, to the nearest tenth

18. _____

Extensions

19. Determine what two whole numbers the square root will be between without using a calculator.
$\sqrt{90}$

19. _____

6.5 Congruent and Similar Triangles

Objective A Decide whether two triangles are congruent.

    Example 1 Determine whether triangle *ABC* is congruent to triangle *XYZ*.

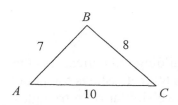
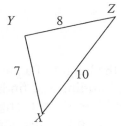

Objective B Find the ratio of corresponding sides in similar triangles.

    Example 2 Find the ratio of corresponding sides for the similar triangles *ABC* and *DEF*.

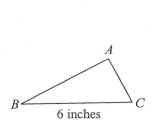
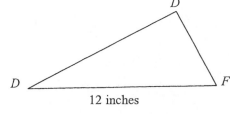

1) yes, 2) $\dfrac{1}{2}$

Examples 6.5 (cont'd)

Objective C Find unknown lengths of sides in similar triangles.

Example 3 Find the missing side x.

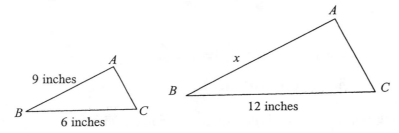

Example 4 Anne needs to find the height of a particular building. She measures the length of the shadow cast by the building and finds that it is 60 feet. She is 5 feet tall and she measures her shadow and it is 12 feet long. Draw two similar right triangles and label them and use the triangles to find the height of the building.

3) 18 inches, 4) 25 feet

Practice Set 6.5

Use the choices below to fill in each blank.

congruent   **similar**

1. Two triangles that have the same shape, but not the same size are called _____.

2. Two triangles that have the same size and same shape are called _____.

Objective A Decide whether the two triangles are congruent.

Are the two triangles congruent?

3.

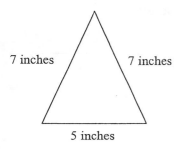   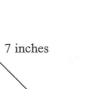

3. _____

4.

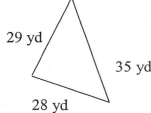   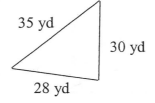

4. _____

5.

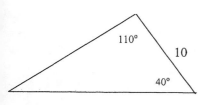   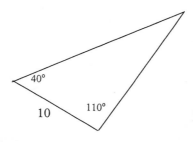

5. _____

Practice Set 6.5 (cont'd)

6.

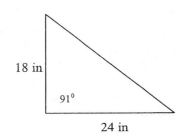

Objective B Find the ratio of corresponding sides in similar triangles.

Find each ratio of the corresponding sides of the given similar triangles.

7.

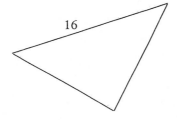

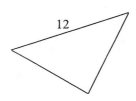

7. _____

8.

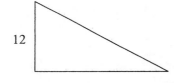

8. _____

9.

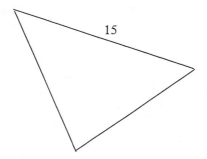

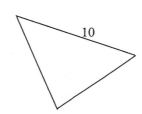

9. _____

Practice Set 6.5 (cont'd)

10.

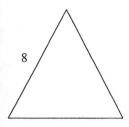

10. _____

Objective C Find the unknown lengths of sides in similar triangles.

Find the unknown length for the given similar triangles.

11.

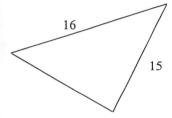

11. _____

12.

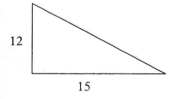 

12. _____

13.

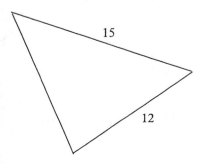 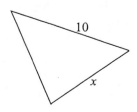

13. _____

Practice Set 6.5 (cont'd)

14.

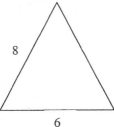

14. _____

15. Give answer to the nearest tenth.

15. _____

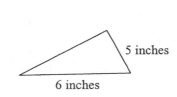 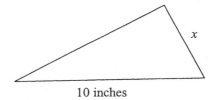

16. Jim needs to know the height of a certain building. He measures the length of the shadow of the building and finds that it is 72 feet. He is six feet tall. His shadow is 9 feet. Find the height of the building.

16. _____

17. Mitch needs to know the height of a certain tree. He measures the length of the shadow of the tree and finds that it is 48 feet. He measures the length of the shadow of a flag pole and find the shadow is 9 feet. If he knows the flag pole is 12 feet high, find the height of the tree.

17. _____

Practice Set 6.5 (cont'd)

Extensions

18. Find the length of the two missing sides of the
    similar triangles.

18.

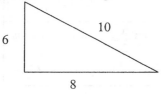 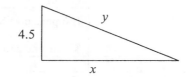

19. Draw two triangles that are similar and show that they are similar by labeling two
    sides and the angle between them.

20. Draw two triangles that are similar and show that they are similar by labeling two
    angles and the side between them.

Chapter 6 Vocabulary Reference Sheet

Term	Definition	Example
	Section 6.1	
Ratio	The quotient of two quantities	The ratio 2 to 5 can be written $\frac{2}{5}$ or 2 : 5.
Unit rate	A rate written with a denominator of 1.	180 miles on 6 gallons can be reduced to $\frac{180}{6} = \frac{30}{1}$, so the unit rate is 30 miles per gallon.
Unit price	Money per item written with a denominator of 1.	If 6 pens cost $2, then $\frac{6}{2} = \frac{3}{1}$ the unit price is $3 for one pen.
	Section 6.2	
Proportion	A statement that two ratios are equal.	$\frac{2}{3} = \frac{4}{6}$ is a proportion.
Cross product	Is used to check if a proportion is true.	$\frac{2}{3} = \frac{4}{6}$ is true since the cross products ($2 \times 6 = 12$ and $3 \times 4 = 12$) are equal.
	Section 6.4	
Perfect square	The result when a number is squared.	1, 4, 9, 16, 25, . . . are perfect squares.
Square root	The square root of a number is the positive amount that was squared to give the number.	$\sqrt{4} = 2$
Radical	Symbol that indicates to find the square root.	$\sqrt{}$ is a radical.
Hypotenuse	The longer side in a right triangle.	The longer side is across from the right angle.
Leg	The shorter two sides of a right triangle.	The shorter sides of a right triangle are adjacent to the right angle.
	Section 6.5	
Congruent	Triangles that are the same size and same shape.	are congruent.
Corresponding	The equal sides or angles of a congruent triangle.	The sides AB and XY are corresponding in the triangles above.
Similar	Triangles that are the same shape but not necessarily the same size.	are similar.

Chapter 6 Practice Test A

Write each ratio or rate as a fraction in lowest terms.

1. 600 cars to 850 cars

1. _____

2. 195 miles in 3 hours

2. _____

3. 1.6 to 4

3. _____

4. $2\frac{1}{3}$ to $1\frac{1}{6}$

4. _____

5. The Oklahoma State University scores 21 points in a football game and the Oklahoma University scores 27 points in the same game. Find the ratio of the score for OSU to OU.

5. _____

Find each unit rate.

6. 270 miles to 6 gallons

6. _____

7. Don runs 10 miles in 2 hours.

7. _____

8. 150 applicants for 10 jobs

8. _____

Find each unit price and decide which is the better buy.

9. Peanut butter:
  10 ounces for $1.58
  16 ounces for $2.06

9. _____

10. Jelly:
  12 ounces for $1.21
  18 ounces for $1.85

10. _____

Chapter 6 Practice Test A (cont'd)

Determine whether the proportion is true.

11. $\dfrac{48}{8} = \dfrac{12}{2}$

11. _____

12. $\dfrac{3.6}{0.4} = \dfrac{8.1}{0.8}$

12. _____

Solve each proportion for the given variable.

13. $\dfrac{x}{4} = \dfrac{9}{6}$

13. _____

14. $\dfrac{3}{x} = \dfrac{4}{6}$

14. _____

15. $\dfrac{3}{\frac{1}{2}} = \dfrac{x}{\frac{2}{3}}$

15. _____

16. $\dfrac{4.9}{7} = \dfrac{2.1}{x}$

16. _____

Chapter 6 Practice Test A (cont'd)

Solve.

17. On a building plan, 2 inches represent 7 feet. How   17. _____
    many inches are needed to represent 17.5 feet?

18. If a car can be driven 195 miles in 3 hours, how long   18. _____
    will it take to travel 325 miles?

19. The standard dose of a certain medicine is 150 ml for   19. _____
    for every 25 pounds of weight. Find the dose for a
    125-pound woman.

Find each square root. Round to the nearest thousandth if necessary.

20. $\sqrt{81}$ 20. _____

21. $\sqrt{163}$ 21. _____

22. $\sqrt{\dfrac{49}{100}}$ 22. _____

Chapter 6 Practice Test A (cont'd)

Solve.

23. Approximate to the nearest hundredth of an inch the . **23.** _____
    Hypotenuse of a right triangle with legs of 5 and 6
    inches.

24. Given that the following triangles are similar, find the   **24.** _____
    unknown length x.

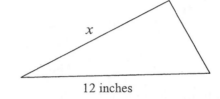

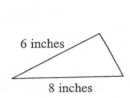

25. A surveyor needs to estimate the height of a tower. He   **25.** _____
    measures the length of his shadow and finds that it is 8
    feet and he knows his height is $5\frac{1}{2}$ feet. He measures
    the length of the shadow of the tower and finds that it is
    48 feet. Draw two right triangles and label them to help
    find the height of the tower.

Chapter 6 Practice Test B

Write each ratio or rate as a fraction in simplest terms.

1. 200 students to 500 students

   a. $\dfrac{200}{500}$ b. $\dfrac{2}{5}$ c. $\dfrac{5}{2}$ d. $\dfrac{1}{2}$

2. 315 miles in 5 hours

   a. $\dfrac{63\text{ miles}}{1\text{ hour}}$ b. $\dfrac{315\text{ miles}}{5\text{ hours}}$ c. $\dfrac{60\text{ miles}}{1\text{ hour}}$ d. $\dfrac{65\text{ miles}}{1\text{ hour}}$

3. 1.2 to 3

   a. 0.4   b. $\dfrac{2}{1}$ c. $\dfrac{1}{8}$ d. $\dfrac{3}{4}$

4. $2\dfrac{1}{3}$ to $1\dfrac{1}{6}$

   a. $\dfrac{1}{2}$ b. $\dfrac{2}{1}$ c. $\dfrac{1}{8}$ d. $\dfrac{1}{4}$

5. The Oklahoma State University scores 21 points in a football game and the Oklahoma University scores 27 points in the same game. Find the ratio of the score for OSU to the total score.

   a. $\dfrac{7}{9}$ b. $\dfrac{21}{27}$ c. $\dfrac{1}{3}$ d. $\dfrac{7}{16}$

Find each unit rate.

6. 175 miles to 7 gallons
   a. 175 miles/7 hours   b. 25 miles/gallon
   c. 30 miles/gallon   d. 28 miles/gallon

7. Pat jogs 12 miles in 2 hours.
   a. 12 miles/2 hours   b. 2 miles/hour
   c. 6 miles/hour   d. 10 miles/hour

Chapter 6 Practice Test B (cont'd)

8. 75 applicants for 15 jobs
  a. 75 applicants/15 jobs
  c. 15 applicants/3 jobs
  b. 25 applicants/5 jobs
  d. 5 applicants/job

Find each unit price and decide which is the better buy.

9. Orange juice:
  64 ounces for $3.49
  32 ounces for $2.08
  a. 64 ounces for $3.49
  b. 32 ounces for $2.08

10. Soda:
  6 cans for $1.29
  24 cans for $5.76
  a. 6 cans for $1.29
  b. 24 cans for $5.76

Determine whether the proportion is true.

11. $\dfrac{7}{8} = \dfrac{54}{60}$
  a. true
  b. false

12. $\dfrac{2.4}{0.6} = \dfrac{3.6}{0.9}$
  a. true
  b. false

Solve each proportion for the given variable.

13. $\dfrac{x}{5} = \dfrac{8}{10}$
  a. 10   **b.** 40   **c.** 6.25   **d.** 4

14. $\dfrac{3.2}{8} = \dfrac{x}{12}$
  a. 2.1   **b.** 38.4   **c.** 4.8   **d.** 1.5

Chapter 6 Practice Test B (cont'd)

15. $\dfrac{\frac{4}{2}}{\frac{2}{3}} = \dfrac{x}{\frac{1}{6}}$

   a. 24   **b.** $\dfrac{1}{4}$ **c.** $\dfrac{2}{3}$ **d.** 1

16. $\dfrac{6}{45} = \dfrac{x}{15}$

   a. 2   **b.** 3   **c.** 6   **d.** 5

Solve.

17. On a building plan, 3.5 inches represent 7 feet. How many feet does 10.5 inches represent?

   a. 9.5 feet   **b.** 21 feet   **c.** 14 feet   **d.** 7 feet

18. If a car can be driven 240 miles in 5 hours, how long will it take to travel 360 miles?

   a. 7.5 hours   **b.** 48 hours   **c.** 9 hours   **d.** 4 hours

19. The standard dose of a certain medicine is 150 ml for every 25 pounds of weight. Find the dose for a 125-pound woman.

   a. 15 ml   **b.** 525 ml   **c.** 750 ml   **d.** 5 ml

Find each square root. Round to the nearest thousandth if necessary.

20. $\sqrt{25}$

   a. $\sqrt{5}$ **b.** 625   **c.** 12   **d.** 5

21. $\sqrt{137}$

   a. 11.704   **b.** 11.705   **c.** 11.7   **d.** 68.5

Chapter 6 Practice Test B (cont'd)

22. $\sqrt{\dfrac{64}{81}}$

     a. $\dfrac{2}{3}$ **b.** $\dfrac{8}{9}$ **c.** $\dfrac{6}{9}$ **d.** $\dfrac{8}{81}$

Solve.

23. Approximate to the nearest hundredth of an in the hypotenuse of a right triangle with legs of 5 and 6 inches.
     a. 400 in   **b.** 20 in   **c.** 7.81 in   **d.** 7.82 in

24. Given that the following triangles are similar, find the unknown length *x*.

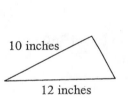 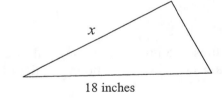

     a. 15 in   **b.** 7 in   **c.** 21.6 in   **d.** 30 in

25. A surveyor needs to estimate the height of a tower. He measures the length of his shadow and finds that it is 9 feet and he knows his height is 6 feet. He measures the length of the shadow of the tower and finds that it is 66 feet. Draw two right triangles and label them to help find the height of the tower.
     a. 100 feet   **b.** 33 feet   **c.** 44 feet   **d.** 72 feet

Name: Date:
Instructor: Section:

7.1 Percents, Decimals, and Fractions

Learning Objectives
- A. Understand percent.
- B. Write percents as decimals or fractions.
- C. Write decimals or fractions as percents.
- D. Applications with percents, decimals, and fractions.

Vocabulary
percent

Objective A Understand percent.

Example 1 In a survey of 100 people, 7 said that they were vegetarians. What percent of the people are vegetarians?

Example 2 In a survey of 100 college students, 73 said that they ate at fast food restaurants at least once a week. What percent of the students eat at a fast food restaurant?

Objective B Write percents as decimals or fractions.

Write each percent as a decimal.

Example 3 15%

Example 4 3.2%

Example 5 175%

Example 6 0.6%

Example 7 200%

Write each percent as a fraction or mixed number in simplest form.

Example 8 30%

Example 9 1.6%

1) 7%, 2) 73%, 3) 0.15, 4) 0.032, 5) 1.75, 6) 0.006, 7) 2, 8) $\frac{3}{10}$, 9) $\frac{2}{125}$

Examples 7.1 (cont'd)

    Example 10 160%

    Example 11 $83\dfrac{1}{3}\%$

    Example 12 300%

Objective C Write decimals as fractions or percents.

    Write each decimal as a percent.

    Example 13 0.63

    Example 14 1.4

    Example 15 0.001

    Example 16 0.7

    Write each fraction or mixed number as a percent.

    Example 17 $\dfrac{3}{20}$

10) $1\dfrac{3}{5}$, *11)* $\dfrac{5}{6}$, *12) 3, 13) 63%, 14) 140%, 15) 0.1%, 16) 70%, 17) 15%*

Examples 7.1 (cont'd)

Example 18 $\dfrac{1}{6}$

Example 19 $1\dfrac{1}{2}$

Example 20 Write $\dfrac{5}{12}$ as a percent. Round to the nearest hundredth percent.

Objective D Applications with percents, decimals, and fractions.

Example 21 46.8% of the students at one junior college are teenagers. Write this as a decimal and as a fraction.

Example 22 An advertisement for a sale reads "everything is $\dfrac{1}{5}$ off". What percent is this?

18) $16\dfrac{2}{3}$%, 19) 150%, 20) 42%, 21) 0.468, $\dfrac{117}{250}$, 22) 20%

Practice Set 7.1

Fill in the blank.

1. Percent means _____.

2. Write the symbol for percent._____

Objective A Understand percent.

3. In a survey of 100 college students, 13 say they drive   3. _____
   more than 10 miles to school. Write this as a percent.

4. In a survey of 100 vehicles in a parking lot, 27 are   4. _____
   pick-ups. What percent are pick-ups?

Objective B Write percents as decimals or fractions.

Write each percent as a decimal.

5. 21%   5. _____

6. 8%   6. _____

7. 3.4%   7. _____

8. 400%   8. _____

Write each percent as a fraction or mixed number in simplest form.

9. 6%   9. _____

10. 42%   10. _____

11. 3.5%   11. _____

Practice Set 7.1 (cont'd)

12. $11\frac{1}{9}\%$

12. _____

Objective C Write decimals or fractions as percents.

Write each decimal as a percent.

13. 0.5

13. _____

14. 0.61

14. _____

15. 0.006

15. _____

16. 4

16. _____

Write each fraction as a percent.

17. $\frac{3}{5}$

17. _____

18. $\frac{17}{20}$

18. _____

19. $3\frac{3}{4}$

19. _____

20. $\frac{1}{6}$

20. _____

21. Write as a percent. Round to the nearest
hundredth percent. $\frac{3}{7}$

21. _____

Practice Set 7.1 (cont'd)

Objective D Applications with percents, decimals, and fractions.

22. In one town, 12% of the population voted in a   22. _____
     election. Write this percent as a fraction in
     simplest form and as a decimal.

Extensions

23. At one college, 30% of the students are freshmen,   23. _____
     26% are sophomores, 25% are juniors, and 15%
     are seniors. What percent are graduate students?

Place < or > between each pair of numbers to make a true statement.

24. 117%   1   24. _____

25. 17%   1   25. _____

7.2 Solving Percent Problems with Equations

Learning Objectives
   A. Write percent problems as equations.
   B. Solve percent problems.

Objective A Write percent problems as equations.

   Translate to an equation.

   Example 1 6 is what percent of 30?

   Example 2 16 is 40% of what number?

   Example 3 What is 17% of 200?

   Example 4 36% of 40 is what?

   Example 5 28% of what number is 700?

   Example 6 What percent of 70 is 20?

Objective B Solve percent problems.

   Example 7 What is 15% of 80?

*1) $6 = 30x$, 2) $16 = 0.4x$, 3) $x = 0.17 \times 200$, 4) $0.36 \times 40 = x$, 5) $0.28x = 700$, 6) $70x = 20$,
7) 12*

Examples 7.2 (cont'd)

Example 8 28% of 600 is what?

Example 9 30% of what is 90?

Example 10 34 is 8.5% of what number?

Example 11 What percent of 16 is 4?

Example 12 48 is what percent of 120?

8) 168, 9) 300, 10) 400, 11) 25%, 12) 40%

Practice Set 7.2

Use the choices below to fill in each blank.
   of   **is**

1. In a stated problem, the equal sign is expressed by the word _____.

2. In a stated problem, multiplication is expressed by the word_____.

Objective A Write percent problems as equations.

Translate each to an equation. Do not solve.

3. 15% of what number is 80?

3. _____

4. 1.7 is 34% of what number?

4. _____

5. What percent of 120 is 4?

5. _____

6. 36 is what percent of 390?

6. _____

7. What number is 10% of 180?

7. _____

8. What number is 56% of 210?

8. _____

9. What percent of 180 is 200?

9. _____

Objective B Solve percent problems.

Solve.

10. 12% of what number is 36?

10. _____

11. What number is 15% of 80?

11. _____

12. 0.99 is 66% of what number?

12. _____

13. 7.5% of what is 10.5?

13. _____

Practice Set 7.2 (cont'd)

14. What percent of 20 is 15?   **14.** _____

15. 38 is what percent of 50?   **15.** _____

16. 648 is 90% of what number?   **16.** _____

17. 0.9 is 3.75% of what number?   **17.** _____

18. 16 is what percent of 25?   **18.** _____

19. 375 is what percent of 750?   **19.** _____

Extensions

20. Write a statement for the equation $0.4x = 16$.   **20.** _____

7.3 Solving Percent Problems Using Proportions

Learning Objectives
  A. Write percent problems as equations.
  B. Solve percent problems.

Vocabulary
amount, base, percent

Objective A Write percent problems as equations.

Translate to an equation.

Example 1 18% of what number is 108?

Example 2 17 is what percent of 85?

Example 3 What number is 70% of 80?

Example 4 179 is 60% of what number?

Example 5 What percent of 90 is 120?

Example 6 80% of 600 is what number?

1) $\dfrac{108}{x} = \dfrac{18}{100}$, 2) $\dfrac{17}{85} = \dfrac{x}{100}$, 3) $\dfrac{x}{80} = \dfrac{70}{100}$, 4) $\dfrac{179}{x} = \dfrac{60}{100}$, 5) $\dfrac{120}{90} = \dfrac{x}{100}$, 6) $\dfrac{x}{600} = \dfrac{80}{100}$

Examples 7.3 (cont'd)

Objective B Solve percent problems.

    Example 7 What number is 70% of 800?

    Example 8 12% of what number is 72?

    Example 9 28.8 is 60% of what number?

    Example 10 What percent of 80 is 16?

    Example 11 180 is what percent of 240?

7) 560, 8) 600, 9) 48, 10) 20%, 11) 75%

Practice Set 7.3

Use the choices below to fill in each blank. The choices will be used more than once.

amount   **base**   **percent**

1. When translating the sentence, 15% of 80 is 12 to a proportion, the number 15 is
    called the _____, the number 80 is called the _____, and
    the number 12 is called the _____.

What part of the percent proportion is unknown in the following statements?

2. What number is 25% of 800? _____

3. 30% of what number is 270? _____

4. 17 is what percent of 25? _____

Objective A Write percent problems as proportions.

Translate each statement to a proportion. Do not solve.

5. 72% of 80 is what number?   5. _____

6. What number is 18% of 25?   6. _____

7. 12.9 is 15% of what number?   7. _____

8. 39% of what number is 112?   8. _____

9. What percent of 490 is 70?   9. _____

Objective B Solve percent problems.

Solve.

10. 25% of 80 is what number?   10. _____

11. 20% of what number is 42?   11. _____

Practice Set 7.3 (cont'd)

12. 80 is what percent of 160?

12. _____

13. 24 is what percent of 12?

13. _____

14. 4.2 is 4% of what number?

14. _____

15. 42 is 28% of what number?

15. _____

16. What number is 37% of 490?

16. _____

17. 120% of what number is 90?

17. _____

18. What percent of 200 is 8?

18. _____

19. 40% of 70 is what number?

19. _____

Extensions

20. Write a word statement for $\dfrac{x}{75} = \dfrac{120}{100}$.

20. _____

7.4 Applications of Percent

Learning Objectives
  A. Solving applications involving percent.
  B. Find percent increase and percent decrease.

Objective A Solve applications involving percent.

Example 1 A community college has an enrollment of 9600. 53% of the students are female. How many females are enrolled?

Example 2 On one snowy day, a high school reported that 256 students out of 421 were late to school. What percent were late? Round to the nearest whole percent.

Example 3 18% of the employees at a factory ride the bus to work. If 81 ride the bus, how many employees work at the factory?

Example 4 A small town reported that it has 8,642 registered voters in 2000 and by 2004 there was a 12% increase in voters.
  a) Find the increase in voters. Round to the nearest whole number.
  b) Find how many registered voters there were in 2004.

1) 5088 females, 2) 61%, 3) 450 employees, 4a) 1037 increase, b) 9679 voters

Examples 7.4 (cont'd)

Objective B Find percent increase and percent decrease.

Example 5 The number of customers at a record store was 15,963 in November of 2006 and was 21,653 in December. Find the percent increase in customers. Round to the nearest whole percent.

Example 6 The number of burglaries in one town was 5624 in 2005 and was 4321 in 2006. Find the percent decrease. Round to the nearest whole percent.

5)36%, 6) 23%

Martin-Gay **Prealgebra** edition 5

Practice Set 7.4

Objective A Solve applications involving percents.

Solve. Round all numbers to the nearest whole number or nearest whole percent.

1. An inspector found 16 defective light bulbs. If this   1. _____
    was 2.5% of the total bulbs inspected, how bulbs
    were inspected?

2. An inspector found that 1.8% of the electronic   2. _____
    components that were tested were defective. If
    2500 components were tested, how many were
    defective?

3. A student's cost for one semester of college was $1850.   3. _____
    The cost for books was $333. What percent of the total
    cost was spent on books?

4. Approximately 40% of one mathematics textbook was   4. _____
    devoted to solving equations. If there were 480 pages
    in the book, how many pages were devoted to equations?

Practice Set 7.4 (cont'd)

5. A local store sold 28 televisions last month. This represented 7% of the yearly sales of televisions. How many televisions were sold for the year?

5. _____

6. A family paid $19,200 for a down payment on a new home. If this was 12% of the total cost of the home, what was the price for the home?

6. _____

7. A student had $90 withheld from his check for taxes. If this was 15% of his total salary, what was his salary before taxes were withheld?

7. _____

Objective B Find percent increase and percent decrease.

Find the amount of increase and the percent increase.

	Original Amount	New Amount	Amount of Increase	Percent Increase
8.	60	75		
9.	20	24		
10.	35	49		

Practice Set 7.4 (cont'd)

Find the amount of decrease and the percent decrease.

	Original Amount	New Amount	Amount of Decrease	Percent Decrease
11.	10	7		
12.	40	24		
13.	150	123		

Solve. Round to the nearest whole percent.

14. A student made a 75 on the first test and a 90 on the   14. _____
    second test. What was the percent increase?

15. The number of students in a college algebra class   15. _____
    decreased from 30 to 24. What was the percent
    decrease?

16. The population of a town increased from 12,600   16. _____
    to 14,490 in five years. What was the percent
    increase?

Practice Set 7.4 (cont'd)

17. The first edition of a math text book had 60 problems on percent. The second edition had 66 problems. Find the percent increase.

17. _____

18. A company had 450 employees, but had to reduce to 300 employees because of budget cuts. Find the percent decrease.

18. _____

19. A hospital had 120 patients in the first week of November. For the second week, there were 138 patients. Find the percent increase.

19. _____

Extensions

20. If a population doubled, what was the percent increase?

20. _____

7.5 Percent and Problem Solving: Sales Tax, Commission, and Discount

---

Learning Objectives
   A. Calculate sales tax and total price.
   B. Calculate commission.
   C. Calculate discount and sale price.

Objective A Calculate sales tax and total price.

Solve. Round all dollar amounts to the nearest cent and percents to the nearest tenth of a percent.

Example 1 Find the sales tax and the total cost on a purchase of $64.95 where there is a 8.5% tax rate.

Example 2 The sales tax on a television that cost $350 was $24.50. What was the sales tax rate?

Objective B Calculate commissions.

Example 3 A real estate broker sold a house for $140,000. If his commission is 2.5% of the selling price, find the amount of the commission.

Example 4 A salesperson earned $1800 for selling $12,000 worth of equipment. Find the commission rate.

Objective C Calculate discount and sale price.

Example 5 Find the discount and the sale price of an item that cost $80 that was on sale for 40% off.

1) $5.52, $70.47, 2) 7%, 3) $3500, 4) 15%, 5) discount $32, sale price $48

Practice Set 7.5

Objective A Calculate sales tax and total price.

Solve each problem. Round all dollar amounts to the nearest cent and all percents to the nearest whole percent.

1. A computer cost $650. What is the tax if the tax rate   1. _____
   is 8.5%?

2. What is the sales tax on a coat that cost $80 if the   2. _____
   tax rate is 6%?

3. The tax on an item was $10.50. If the tax rate is 7%,   3. _____
   find the original cost of the item and the total cost
   with tax.

4. A gold bracelet sells for $600. What is the tax and the   4. _____
   total cost if there is a 5% tax rate?

5. The tax on a new chair was $27. Find the cost of the   5. _____
   chair and the total cost if the tax rate is 6%.

6. A wedding dress sells for $360. Find the total cost   6. _____
   if the tax rate is 5.5%.

Practice Set 7.5 (cont'd)

7. The tax on a new car was $1425. If the tax rate 7.5%, what was the price of the car?

7. _____

8. A cell phone cost $49. If the tax rate was 8%, find total cost of the phone plus tax.

8. _____

Objective B Calculate commissions.

9. A home sold for $750,000. If the realtor receives a 6% commission, how much will the commission be?

9. _____

10. A salesman made $800,000 worth of sales in new cars. If he gets 9% commission, what will his commission be?

10. _____

11. A sales clerk made $840 in commission one month. If she gets a 7% commission on her sales, how much were her sales that month?

11. _____

12. A realtor made $7200 commission on the sale of a house. If the commission is 4% of the price of the house, what was the price of the house?

12. _____

Practice Set 7.5 (cont'd)

Objective B Calculate discount and sale price.

	Original Price	Discount Rate	Amount of Discount	Sale Price
13.	$92	8%		
14.	$48	15%		
15.	$150	30%		
16.	$650	20%		

17. A television that is regularly $400 is on sale for   17. _____
    for 25% off. Find the sale price.

18. A $750 dinning room set is on sale for 20% off.   18. _____
    What is the discount and the sale price?

Extensions

19. A CD player that normally costs $60 is on sale   19. _____
    for 15% off. If there is a 6% tax, find the cost
    after the discount has been deducted the tax has
    been added.

20. Which is better: A 25% discount followed by   20. _____
    a 10% discount or a 30% discount followed by a
    5% discount?

7.6 Percent and Problem Solving: Interest

Learning Objectives
    A. Calculate simple interest.
    B. Calculate compound interest.

Vocabulary
interest, principal, simple interest, compound interest

Objective A Calculate simple interest.

    Example 1 Find the simple interest on $1000 for 3 years at 8%.

    Example 2 Find the simple interest on $4200 for 9 months at 12%.

    Example 3 Find the total amount of money after 3 years if $3000 is invested at 15% simple interest.

Objective B Calculate compound interest.

    Example 4 $2500 is invested at 4.8% compounded annually for 8 years. Find the total amount at the end of 8 years.

    Example 5 $8000 is invested at 7.4% compounded quarterly for 7 years. Find the total amount of money after 7 years.

1) $240, 2) $378, 3) $4350, 4) $3637.73, 5) $13,365.92

Practice Set 7.6

Use the choices below to fill in each blank. Choices will be used more than once.

   compound   simple

1. The formula $I = Prt$ is used to find _____ interest.

2. The formula $A = P\left(1 + \dfrac{r}{n}\right)^{nt}$ is used to find _____ interest.

3. If interest is computed not only on the original principle but also on the interest already earned it is called _____ interest.

4. If interest is computed only on the original principle, it is called _____ interest.

Objective A Calculate simple interest.

Find the simple interest and the total amount of money at the end of the time.

	Principal	Rate	Time	Interest	Total Amount
5.	$800	9%	3 years		
6.	$700	8%	8 years		
7.	$1200	6.5%	4 years		
8.	$550	8%	9 months		
9.	$2800	5%	6 months		

10. A company borrows $10,000 for 6 years on a simple interest loan at 12%. Find the interest paid on the loan.

    10. _____

11. A bank advertises a CD with a simple interest of 5% for 3 years. If you buy a CD for $1200, how much interest will you make after 3 years? What will be your total amount after 3 years?

    11. _____

Practice Set 7.6 (cont'd)

12. If you borrow $3000 for a used car from a friend   12. _____
    and agree to pay it all back in one year with simple
    interest of 6%, how much interest will you owe?
    What will be the total amount you owe?

Objective B Calculate compound interest.

	Principal	Interest Rate	Compounded	Time	Total Amount
13.	$12,000	8%	annually	4 years	
14.	$8000	12%	monthly	6 years	
15.	$3000	15%	quarterly	3 years	
16.	$24,000	8%	semiannually	5 years	
17.	$300	9%	quarterly	20 years	
18.	$500	3%	daily	8 years	

Extensions

19. If $8000 is invested at 12% compounded   19. _____
    quarterly for 8 years, how much interest is
    made?

20. Which is better: 20. _____
    $2000 at 8% semiannually for 5 years
    $2000 at 7.5% quarterly for 5 years
    $2000 at 12% simple interest for 5 years

Chapter 7 Vocabulary Reference Sheet

Term	Definition	Example
	Section 7.1	
Percent	Per hundred	29% is 29 per hundred or 29 hundredths
	Section 7.3	
Amount	Part compared to a whole.	In "250 is what percent of 500", 250 is the amount
Base	Appears after *of* in a percent problem, it is the whole.	In "250 is what percent of 500", 500 is the base
Percent	%	In "250 is what percent of 500", The percent is to be solved for.
	Section 7.6	
Interest	Money earned on an investment.	In "$2000 is invested at 12% for 6 years", 12% is the interest
Principal	The original amount of money in an investment.	In "$2000 is invested at 12% for 6 years", $2000 is the principal.
Simple Interest	Interest earned only on the principal.	The formula for simple interest is $I = Prt$.
Compound Interest	Interest earned on principal and on interest already earned.	The formula for compound interest is $A = P\left(1 + \dfrac{r}{n}\right)^{nt}$.

Martin-Gay **Prealgebra** edition 5

Chapter 7 Practice Test A

Write each percent as a decimal.

1. 32%   **1.** _____

2. 150%   **2.** _____

3. 0.05%   **3.** _____

Write each decimal as a percent.

4. 0.6   **4.** _____

5. 3.2   **5.** _____

6. 0.012   **6.** _____

Write each percent as a fraction or a mixed number in simplest form.

7. 36%   **7.** _____

8. 140%   **8.** _____

9. 0.04%   **9.** _____

Write each fraction or mixed number as a percent.

10. $\dfrac{7}{25}$ **10.** _____

11. $\dfrac{17}{20}$ **11.** _____

12. $1\dfrac{5}{8}$ **12.** _____

13. Enrollment has increased by $\dfrac{1}{4}$ over ten years. Write   **13.** _____

$\dfrac{1}{4}$ as a percent.

Chapter 7 Practice Test A (cont'd)

14. The number of small businesses has increased by 8%.   14. _____
    Write 8% as a fraction in simplest form.

Solve.

15. What number is 28% of 600?   15. _____

16. 0.5% of what number is 10?   16. _____

17. 192 is what percent of 768?   17. _____

Solve. If necessary round percents to the nearest tenth, dollar amounts to the nearest cent, and all other numbers to the nearest whole.

18. A saline solution is 0.9% salt. How much salt is in 1000 ml?   18. _____

19. A cotton farmer in Texas lost 35% of his potential crop   19. _____
    to hail. If he estimated that he lost $87,500, find the
    total value of his crop.

20. If the local tax rate is 6.5%, find the total amount charged   20. _____
    for a stereo that is priced at $580.

21. A town's population increased from 37,900 to 42,448.   21. _____
    Find the percent increase.

Chapter 7 Practice Test A (cont'd)

22. A flat screen monitor is on sale for 25% off. What would the sale price be if the original price was $360?

22. _____

23. A sales clerk is paid 6% commission on all sales. Find the commission on $12,000 in sales.

23. _____

24. A sales tax of $13.19 is added to an item's price of $164.88. What is the sales tax rate?

24. _____

25. Find the simple interest earned on $3000 saved for $2\frac{1}{2}$ years at 9% interest.

25. _____

26. $1500 is compounded annually at 8% for 6 years. Find the total amount after 6 years.

26. _____

27. A student borrowed $1200 from a bank at 12.5% compounded quarterly for 2 years. Find the total owed after 2 years.

27. _____

Name:
Instructor:

Date:
Section:

Chapter 7 Practice Test B

Write each percent as a decimal.

1. 28%
    a. 28   **b.** 0.28   **c.** 2.8   **d.** 2800

2. 124%
    a. 124   **b.** 0.124   **c.** 1.24   **d.** 12,400

3. 0.6%
    a. 0.6   **b.** 60   **c.** 0.06   **d.** 0.006

Write each decimal as a percent.

4. 0.7
    a. 7%   **b.** 0.7%   **c.** 70%   **d.** 0.007%

5. 4.1
    a. 410%   **b.** 41%   **c.** 4.1%   **d.** 0.041%

6. 0.028
    a. 0.028%   **b.** 2.8%   **c.** 28%   **d.** 280%

Write each percent as a fraction or a mixed number in simplest form.

7. 42%
    a. $\frac{21}{50}$ **b.** 42   **c.** $\frac{42}{100}$ **d.** $4\frac{1}{5}$

8. 160%
    a. $\frac{7}{50}$ **b.** $\frac{4}{25}$ **c.** $1\frac{3}{5}$ **d.** $1\frac{4}{25}$

9. 0.3%
    a. $\frac{3}{10}$ **b.** $\frac{3}{100}$ **c.** 3   **d.** $\frac{3}{1000}$

Martin-Gay **Prealgebra** edition 5

270

Chapter 7 Practice Test B (cont'd)

Write each fraction or mixed number as a percent.

10. $\dfrac{7}{10}$

   a. 7%   **b.** 70%   **c.** 0.7%   **d.** 0.07%

11. $\dfrac{3}{4}$

   a. 75%   **b.** 0.75%   **c.** 3.4%   **d.** 340%

12. $2\dfrac{1}{5}$

   a. 2.2%   **b.** 0.22%   **c.** 220%   **d.** 22%

13. Enrollment has increased by $\dfrac{3}{5}$ over ten years. Write $\dfrac{3}{5}$ as a percent.

   a. 60%   **b.** 0.6%   **c.** 35%   **d.** 3.5%

14. The number of employees at a small business has increased by 12%. Write 12% as a fraction in simplest form.

   a. 12   **b.** $\dfrac{3}{25}$ **c.** $1\dfrac{1}{5}$ **d.** $\dfrac{1}{2}$

Solve.

15. What number is 16% of 400?
   a. 25   **b.** 6400   **c.** 2500   **d.** 64

16. 14% of what number is 49?

   a. 350   **b.** $\dfrac{2}{7}$ **c.** 35   **d.** 6.86

17. 180 is what percent of 900?
   a. 5%   **b.** 50%   **c.** 20%   **d.** 0.2%

Chapter 7 Practice Set B (cont'd)

Solve. If necessary round percents to the nearest tenth, dollar amounts to the nearest cent, and all other numbers to the nearest whole.

18. A silver bracelet is 80% silver. If the bracelet weighs 6 g, how much of the weight is silver?

    a. 7.5 g   **b.** 4.8 g   **c.** 48 g   **d.** 1.3 g

19. A cotton farmer in Texas lost 20% of his potential crop to hail. If he estimated that he lost $42,000, find the total value of his crop.

    a. $210,000   **b.** $8400   **c.** $8,400,000 **d.** $21,000

20. If the local tax rate is 8%, find the total amount charged for a stereo that is priced at $650.

    a. $520   **b.** $598   **c.** $52   **d.** $702

21. A town's population increased from 65,892 to 69,421. Find the percent increase.

    a. 95%   **b.** 5.1%   **c.** 5.4%   **d.** 5%

22. A flat screen monitor is on sale for 20% off. What would the sale price be if the original price was $250?

    a. $200   **b.** $50   **c.** $230   **d.** $20

23. A sales clerk is paid 10% commission on all sales. Find the commission on $17,000 in sales.

    a. $170,000   **b.** $1700   **c.** $17   **d.** $170

24. A sales tax of $5.78 is added to an item's price of $96.40. What is the sales tax rate?

    a. 1%   **b.** 5%   **c.** 6%   **d.** 17%

25. Find the simple interest earned on $6000 saved for 3 years at 8% interest.

    a. $144,000   **b.** $7,440,000   **c.** $7440   **d.** $1440

26. $8000 is compounded annually at 10% for 3 years. Find the total amount after 3 years.

    a. $10,648   **b.** $70,400   **c.** $2400   **d.** $24,000

27. A student borrowed $6000 from a bank at 8% compounded quarterly for 4 years. Find the total owed after 4 years.

    a. $7920   **b.** $8,236.71   **c.** $1920   **d.** $8162.93

8.1 Reading Pictographs, Bar Graphs, Histograms, and Line Graphs

Learning Objectives
  A. Read pictograms.
  B. Read and construct bar graphs.
  C. Read and construct histograms.
  D. Read line graphs.

Vocabulary
Pictograms, bar graphs, histograms, line graphs

Objective A Read pictographs.

Example 1 Use the pictograph to answer the following questions.

Peanut Production in 2005	🥜 = $50 million
Georgia	🥜🥜🥜🥜🥜🥜🥜
Texas	🥜🥜🥜
Alabama	🥜🥜
Florida	🥜 🥜
North Carolina	🥜

  (a) What state had the most peanut production?

  (b) How much money was made in Texas in peanut production?

Objective B Read and construct bar graphs.

Example 2 Use the following bar graph to answer the questions.

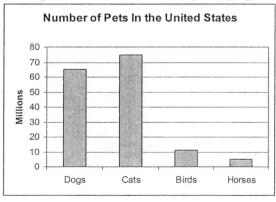

  (a) Estimate the number of dogs.

  (b) What type of pet is the most popular?

1a) Georgia, b) $150,000,000 2a) 65,000,000 dogs b) cat

Examples 8.1 (cont'd)

Example 3 Draw a vertical bar graph using the information in the table.

	Calories
Apple	47
Banana	95
Orange	37
Pear	40

Objective C Read and construct histograms.

Use the following histogram to answer the following question.

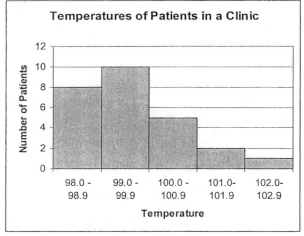

Example 4 How many patients had a temperature between 99.0 and 99.9?

Example 5 How many patients had a temperature of 101.0 or above?

3) 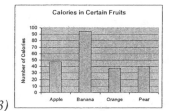   *4) 10 patients, 5) 3 patients*

Examples 8.1 (cont'd)

Use the following data for Examples 6 and 7.
Grades on a Quiz in Statistics

76	84	98	62	98
94	83	65	78	85
86	75	83	99	84
76	78	90	84	62

Example 6 Complete the frequency distribution for the data on grades.

Class Intervals (Grades)	Tally	Class Frequency (Number of Grades)
60 - 69		
70 - 79		
80 - 89		
90 - 99		

Example 7 Construct a histogram from the frequency distribution table in example 6.

6)

Class Intervals (Grades)	Tally	Class Frequency (Number of Grades)
60 - 69	///	3
70 - 79	/////	5
80 - 89	///////	7
90 - 99	/////	5

7)

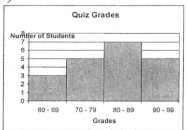

Examples 8.1 (cont'd)

Objective D Read line graphs.

Example 8 Use the line graph to answer the following questions.

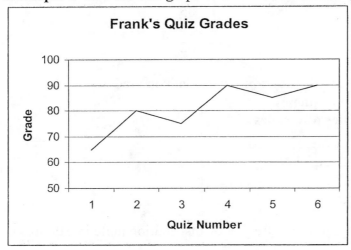

(a) What was Frank's highest quiz grade?

(b) What did Frank make on Quiz 5?

(c) On which quiz did Frank make 80?

8a) 90, b) 85, c) Quiz 2

Practice Set 8.1

Use the choices below to fill in each blank.

| bar | class frequency | class interval |
| histogram | line | pictograph |

1. A graph that uses pictures or symbols to present data is called a

   _____.

2. A graph that displays information that connects dots is called a
   _____ graph.

3. A graph that gives data in horizontal or vertical bars is called a
   _____ graph.

4. A _____ is a bar graph where the width of each bar
   represents a _____ and the height of the bar represents
   the _____.

Objective A Read pictographs.

Use the following pictograph to answer exercises 5 - 8.

Cotton Production in 2005   = $100 million

Texas
California
Georgia
Mississippi
Arkansas

5. What state produced the most money from cotton?   5. _____

6. What state produced $900,000,000 in cotton?   6. _____

7. Georgia produced how much cotton?   7. _____

Practice Set 8.1 (cont'd)

Objective B Read and construct bar graphs.

Use the following bar graph to answer exercises 8 - 10.

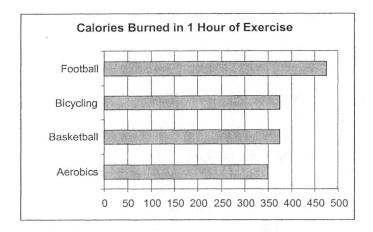

8. What two activities burn the same calories?   8. _____

9. How many calories are burned in one hour of football?   9. _____

10. How many more calories are burned by bicycling   10. _____
    than by aerobics?

11. Construct a bar graph from the following information.
    Population for the 5 largest cities in Texas in 2005.

City	Population (in thousands)
Houston	2,954
Dallas	1,189
San Antonio	1,145
El Paso	564

Exercises 8.1 (cont'd)

Objective C Read and construct histograms.

Use the following histogram for exercises 12 - 14.

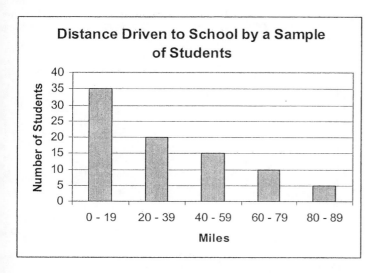

12. How many students drove between 20 and 39 miles?   12. _____

13. Estimate how many students drove between 40   13. _____
    and 59 miles.

14. How many students drove more than 60 miles?   14. _____

15. Complete the frequency distribution table for the data on years of service.
    Years of Service for Employees at Richardson's Factory

3	5	2	1	8	10	8
12	8	5	9	13	20	3
18	21	14	17	8	9	

Class Intervals (Years)	Tally	Class Frequency (Number of Employees)
0 - 4		
5 - 9		
10 - 14		
15 - 19		
20 - 24		

Exercises 8.1 (cont'd)

16. Construct a histogram from the frequency distribution table in example 15.

Objective D Read line graphs.

Use the following line graph for exercises 17 - 18.

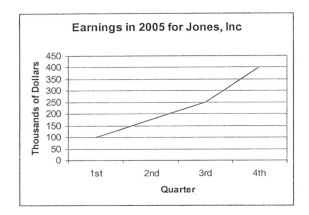

17. What were the earnings for the first quarter?   17. _____

18. What quarter earned the most?   18. _____

Extensions

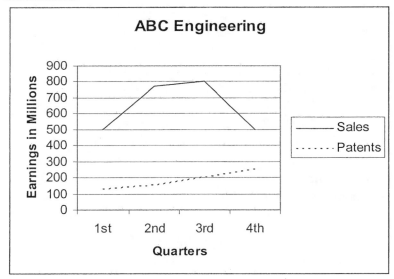

19. How much did the company make in the third quarter   19. _____
on patents?

20. How much more did the company earn in the 3rd 20. _____
quarter in sales than in patents?

8.2 Reading Circle Graphs

Learning Objectives
   A. Read circle graphs.
   B. Draw circle graphs.

Vocabulary
circle graph

Objective A Read a circle graph.

Use the circle graph to answer the questions.

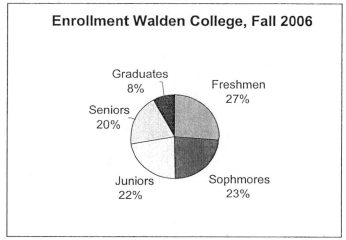

Example 1 What percent are freshmen?

Example 2 What percent are undergraduates?

Example 3 If the total enrollment at the school was 17,100, find the number of freshmen enrolled.

1) 27%, 2) 92%, 3) 4617 freshmen

Examples 8.2 (cont'd)

Objective B Draw circle graphs.

Example 4 Draw a circle graph using the following information.

Grade in Elementary Statistics	Percent
A	25
B	30
C	30
D	10
F	5

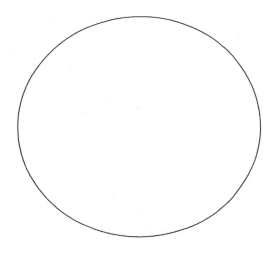

5)

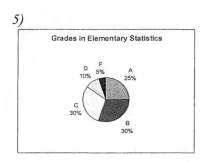

Practice Set 8.2

Fill in each blank.

1. The total number of degrees in a circle is _____.

2. The percents in a circle graph should add up to _____.

Objective A Read circle graphs.

Use the graph to answer questions 3 - 7.

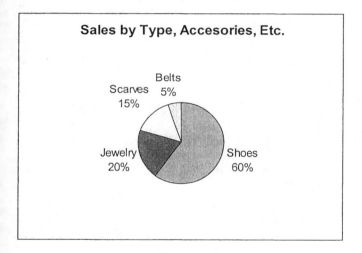

Sales by Type, Accesories, Etc.

Belts 5%
Scarves 15%
Jewelry 20%
Shoes 60%

3. What item sold the most?   3. _____

4. What is the ratio of jewelry to shoes?   4. _____

5. What is the ratio of all other sales to shoes?   5. _____

6. What is the ratio of belts to scarves?   6. _____

7. What is the ratio of scarves to jewelry?   7. _____

Practice Set 8.2 (cont'd)

Use the graph to answer questions 8 - 12.

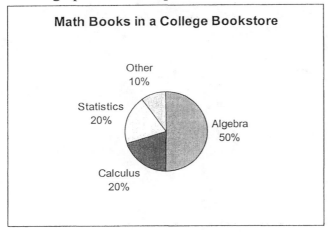

8. What represents the largest type of books?   8. _____

9. What present is accounted for by Calculus and   9. _____
    Statistics?

10. What two types of books had the same percentage?   10. _____

11. If there are 600 math books, how many are algebra? 11. _____

12. How many are statistics books?   12. _____

Use the graph for questions 13 - 17.

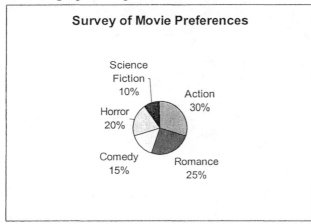

13. What percent prefer action movies?   13. _____

14. What is the total percent of science fiction and   14. _____
    horror?

Practice Set 8.2 (cont'd)

15. What is the most preferred type of movie?

15. _____

16. What is the second preferred type of movie?

16. _____

17. What is the ratio of horror to action?

17. _____

Objective B Draw circle graphs.

Make a circle graph for each of the following sets of data.

18.

Quarterly Sales	Percent	Degrees
1st Quarter	15%	
2nd Quarter	15%	
3rd Quarter	30%	
4th Quarter	40%	

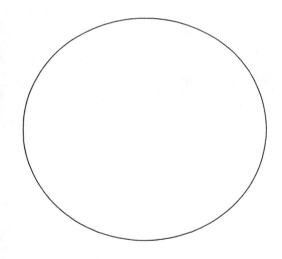

Practice Set 8.2 (cont'd)

19.

	Percent	Degrees
Housing	17%	
Transportation	16%	
Food	28%	
Entertainment	17%	
Clothes	20%	
Other	2%	

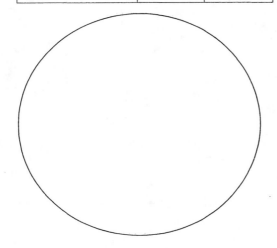

Concept Extensions
Make a circle graph for the following information. Acres per crop in Gaines County, Texas.

20.

Crop	Acres	Percent	Degrees
Peanuts	264,000		
Cotton	66,400		
Wheat	7300		

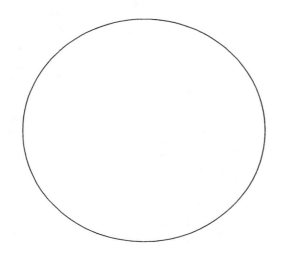

8.3 The Rectangular Coordinate System and Paired Data

Learning Objectives
  A. Plot points on a rectangular coordinate system.
  B. Determine whether ordered pairs are solutions of equations.
  C. Complete ordered-pair solutions of equations.

Vocabulary
plane, paired data, rectangular coordinate system, origin, x-axis, y-axis, linear equation

---

Objective A Plot points on a rectangular coordinate system.

Example 1 Plot each point corresponding to the ordered pairs on the same set of axes.

$(3, 2), (-1, 2), (0, 4), (-3, -4), (3, -2), (-3, 0)$

Example 2 Find the ordered pair corresponding to each point plotted below.

1)

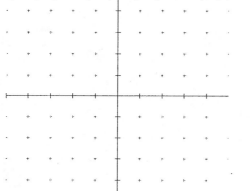

2) A(4, 2), B(1, 0), C(-3, 2), D(0, -2), E(-2, -3)

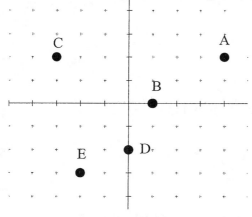

Martin-Gay **Prealgebra** edition 5   287

Examples 8.3 (cont'd)

Objective B Determine whether ordered pairs are solutions of equations.

Example 3 Is (2, 3) a solution of $2x - y = 1$?

Example 4 Each ordered pair listed is a solution of the equation $x - y = 2$. Plot them
on the same set of axes.

     a) (2, 0)   b) (0, −2)   c) (4, 2)   d) (3, 1)   e) (−1, −3)

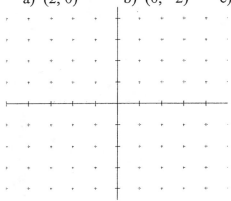

Objective C Complete ordered-pair solutions.

Example 5 Complete each ordered-pair solution of the equation $y = x + 2$.
    a) (2,)   b) (, 0)   c) (4,)

Example 6 Complete each ordered-pair solution of the equation $y = 2x - 3$.
    a) (4,)   b) (, 7)

3) yes, 4)   *5a) (2, 4), b) (−2, 0), c) (4, 6), 6a) (4, 5), b) (5, 7)*

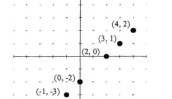

Practice Set 8.3

Use the choices below to fill in each blank.

    x   *y*   **origin**   **quadrants**   **(0, 0)**   **plane**

1. In the ordered pair (4, −3), the 4 is the _____-value and the −3 is the
    _____-value.

2. The point where the *x*-axis and the *y*-axis cross is called the _____.

3. The axes divide the plane into four regions called _____.

4. The origin is the ordered pair _____.

5. A flat surface that extends indefinitely in all directions is called a _____.

Objective A Plot points on a rectangular coordinate system.

Plot the points corresponding to the ordered pairs.

6. (3, 4), (1, 2), (−2, 3), (−2, −1), (0, 4)

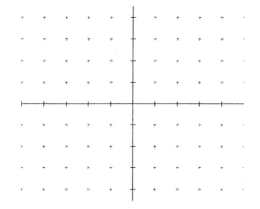

Practice Set 8.3 (cont'd)

7. (1, 4), (−3, 4), (2, −1), (2, 0), (0, 2)

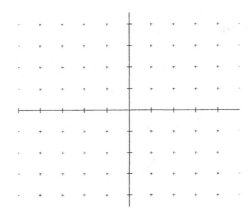

Find the *x*- and *y*- coordinates of each labeled point.

8.

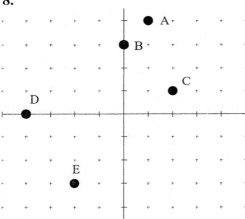

8. _____

9.

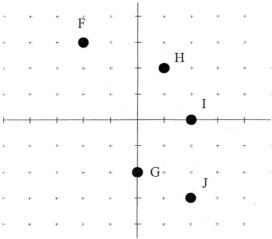

9. _____

Practice Set 8.3 (cont'd)

Objective B Determine whether ordered pairs are solutions of equations.

Determine if the ordered pair is a solution to the given linear equation.

10. $(1, 4)$, $y = 4x$ **10.** _____

11. $(2, 2)$, $x - y = 4$ **11.** _____

12. $(5, 2)$, $x = 3y - 1$ **12.** _____

13. $(4, 1)$, $2x - 3y = 4$ **13.** _____

Use the rectangular coordinate system to plot the three ordered-pair solutions of the given equation.

14. $x + y = 3$, $(1, 2)$, $(3, 0)$, $(0, 3)$

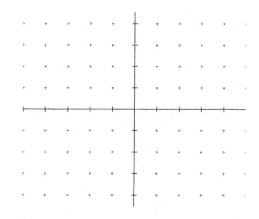

Practice Set 8.3 (cont'd)

15. $x = -3y$, $(3, -1)$, $(0, 0)$, $(-3, 1)$

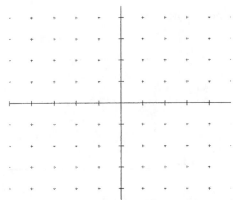

Objective C Complete ordered-pair solutions of equations.

Complete each ordered-pair solution for the given equations.

16. $y = -4x$, $(1, \ )$, $(-2, \ )$, $(\ , 0)$ 16. _____

17. $2x - y = 4$, $(3, \ )$, $(\ , 0)$, $(4, \ )$ 17. _____

18. $x + y = 0$, $(-2, \ )$, $(4, \ )$, $(0, \ )$ 18. _____

19. $y = x - 3$, $(4, \ )$, $(5, \ )$, $(\ , -3)$ 19. _____

Extensions

20. Tell which quadrant each point lies in.   20. _____
      (a) $(-3, 2)$ (b) $(2, 3)$ (c) $(-3, -6)$ (d) $(3, -5)$

8.4 Graphing Linear Equations in Two Variables

Learning Objectives
   A. Graph linear equations by plotting points.

Objective A Graph linear equations by plotting points.

Example 1 Graph the equation $y = 2x$ by plotting the following points that satisfy the equation and drawing a line through the points.
(1, 2), (2, 4), (0, 0)

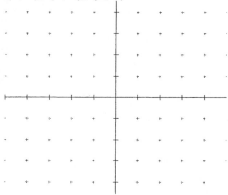

Example 2 Graph $x - y = 4$.   **Example 3** Graph $y = 2x - 1$.

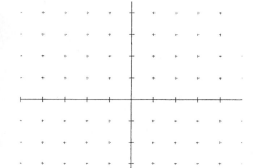

1)   2)   3)

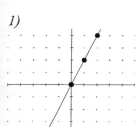 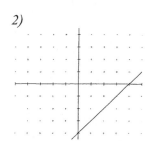 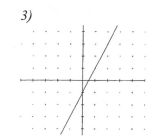

Example 8.4 (cont'd)

Example 4 Graph $x = 2$.

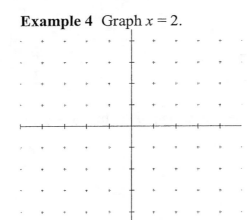

Example 5 Graph $y = -3$.

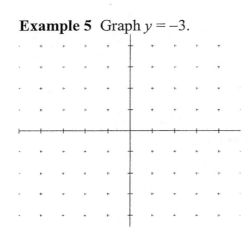

4)   5)

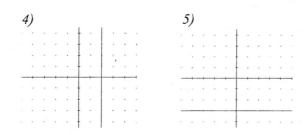

Name:

Instructor:

Date:

Section:

Practice Set 8.4

Use the choices below to fill in each blank.

    vertical   **horizontal**

1. The graph of the equation $y = 4$ would be a _____ line.

2. The graph of the equation $x = 2$ would be a _____ line.

Objective A Graph linear equations by plotting points.

Graph each equation.

3. $x + y = 5$

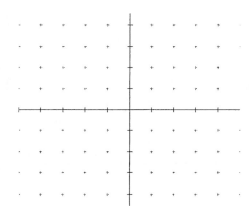

4. $x - y = 3$

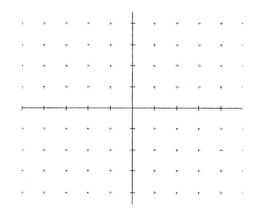

5. $y = 2x - 2$

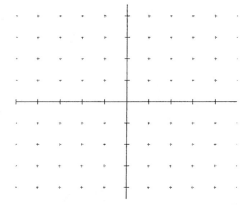

6. $2x + y = 4$

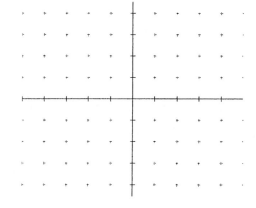

Practice Set 8.4 (cont'd)

7. $y = -2x + 1$

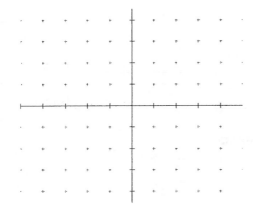

8. $x = y + 1$

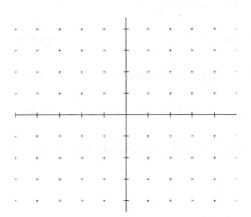

9. $x = 3y$

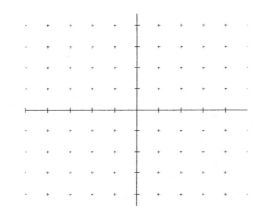

10. $y = x - 1$

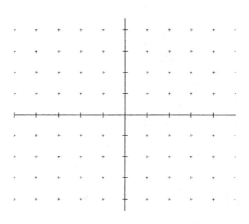

11. $x = 2$

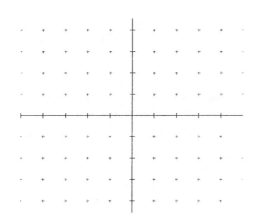

12. $y = 3$

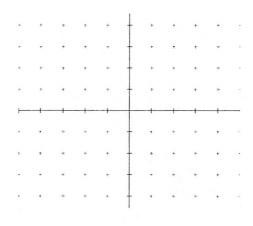

Practice Set 8.4 (cont'd)

13. $x = -3$

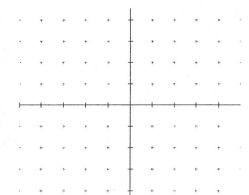

14. $y = -2$

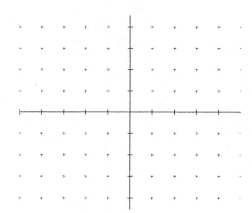

15. $x = y$

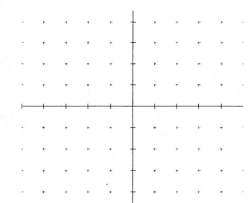

16. $2x + y = 3$

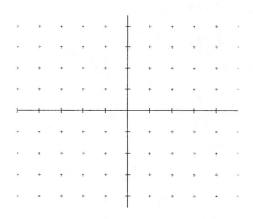

17. $y = \dfrac{1}{2}x$

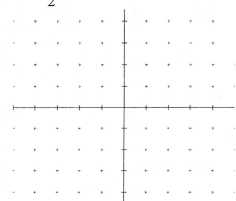

18. $x + 2y = 2$

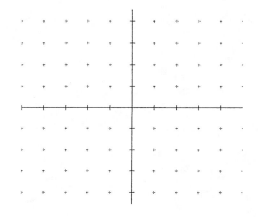

Practice Set 8.4 (cont'd)

19. $y = \dfrac{1}{3}x - 1$

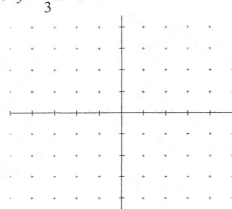

20. Fill in the table and use it to graph $y = |x + 2|$.

x	y
−4	
−3	
−2	
−1	
0	
1	

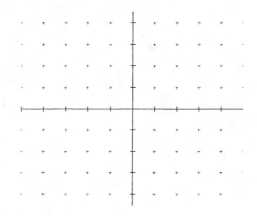

8.5 Counting and Introduction to Probability

Learning Objectives
  A. Use a tree diagram.
  B. Find the probability of an event.

Vocabulary
probability, outcome

Objective A Use a tree diagram.

Example 1 Draw a tree diagram for having two children.

Example 2 Draw a tree diagram for answering two true-false questions.

1)   *2)*

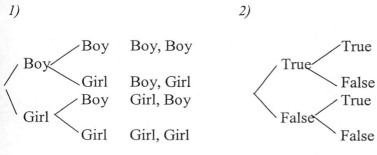

Examples 8.5 (cont'd)

Objective B Find the probability of an event.

Example 3 If a couple has 2 children, find the probability of getting 2 girls.

Example 4 If a die is rolled one time, find the probability of getting a number less than 2.

Example 5 Find the probability of choosing a white marble from a box containing 3 red, 1 blue, and 2 white marbles.

3) $\frac{1}{4}$, 4) $\frac{1}{6}$, 5) $\frac{1}{3}$

Practice Set 8.5

Use the choices below to fill in each blank. Some choices will be used more than once.

> **0**   **1**   **probability**

1. The measure of the likelihood of an event occurring is called the

_____.

2. _____ is calculated by dividing the number of ways that an event can occur by the total number of outcomes.

3. A probability can never be larger than _____.

4. A probability can never be smaller than _____.

Objective A Use a tree diagram to count outcomes.

5. Use a tree diagram to find the possible outcomes of choosing a number from 1 or 2 and then choosing a letter from the letters "of".

6. Use a tree diagram to find the possible outcomes if you pick a number from 1 or 2 and then pick a letter a or b.

7. Use a tree diagram to find the possible outcomes of flipping a coin 3 times.

Practice Set 8.5 (cont'd)

Objective B Find the probability of an event.

If a die is rolled one time, find the following probabilities.

8. a 6   8. _____

9. a 2 or a 3   9. _____

10. a number less than 4   10. _____

11. a number greater than 8   11. _____

One marble is drawn from a box with 4 red marbles, 3 blue marbles, and 6 white marbles. Find the following probabilities.

12. a red marble   12. _____

13. a blue marble   13. _____

14. a white marble   14. _____

15. a white or a blue marble   15. _____

Twenty people were on the same diet. If 12 people lost weight, 3 people stayed the same weight, and 5 people gained weight, find the following probabilities.

16. a person lost weight   16. _____

17. a person gained weight   17. _____

18. a person stayed the same weight   18. _____

Extensions

19. If you flip a coin 3 times , find the probability   19. _____
that you will get all heads.

20. If you flip a coin 3 times, find the probability   20. _____
that you will get 1 head and 2 tails, in any order.

Chapter 8 Vocabulary Reference Sheet

Term	Definition	Example
	Section 8.1	
Pictogram	A graph where pictures or symbols are used to represent data.	might represent 100 kittens.
Bar Graph	A graph where horizontal or vertical bars are used to represent data.	
Histogram	A bar graph where the width of each bar represents a range of numbers.	
Line Graph	Displays information with a line that connects data points.	
	Section 8.2	
Circle Graphs	A graph where each section of the circle represents a category and the relative size of that category.	
	Section 8.3	
Plane	A flat surface that extends indefinitely.	
Paired data	A relationship between two quantities.	(2, 3)

Rectangular Coordinate System	Is used to describe location of points on a plane by using an horizontal and a vertical number line.	
x-axis	The horizontal number line on a rectangular coordinate system.	
y-axis	The vertical number line on a rectangular coordinate system.	
Origin	The intersection of the x-axis and y-axis on a rectangular coordinate system.	
Linear equation	An equation with x and y variables with exponents of 1.	$2x + 3y = 5$ is a linear equation.
8.5		
Outcome	A possible result of an experiment.	If you flip a coin one time, head is one possible outcome.
Probability	The measure of the chance of an event occurring.	If you flip a coin one time, the probability of getting a head is $\frac{1}{2}$.

Chapter 8 Practice Test A

The following pictograph shows the number of puppies one pet store sold the first three months of 2006. Use this graph to answer questions 1 - 3.

Month	Puppies Sold
January	
February	
March	

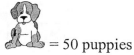 = 50 puppies

1. In what month were the most puppies sold?   1. _____

2. How many puppies were sold in February?   2. _____

3. How many more puppies were sold in January then in March?   3. _____

The bar graph shows the number of moviegoers at a theater for each week in May. Use the bar graph to answer questions 4 - 6.

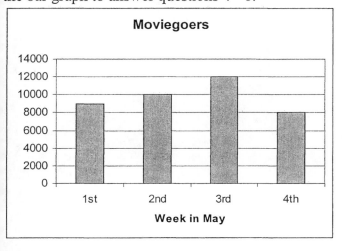

4. How many weeks had 10,000 or more moviegoers?   4. _____

5. How many moviegoers were there in the 4th week?   5. _____

6. Which week had the fewest moviegoers?   6. _____

Chapter 8 Practice Test A

7. Use the following information to draw a bar graph.

Country	Life Expectancy
Japan	81
Canada	80
United Kingdom	79
United States	78
Mexico	75

The following line graph shows the enrollment for Williams Community College. Use the graph to answer questions 8 - 10.

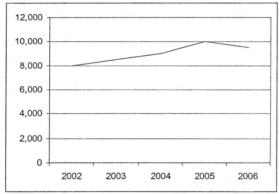

8. What was the enrollment in 2005?   8. _____

9. Between what two years was there a decrease in enrollment?   9. _____

10. What was the enrollment in 2002?   10. _____

Chapter 8 Practice Test A

The circle graph shows the result of a survey of what type of vehicle is being driven. Use the graph to answer questions 11 - 14.

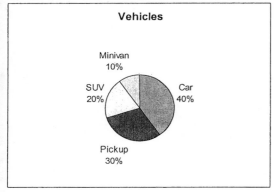

11. Find the ratio of pickups to cars.   11. _____

12. Find the ratio of SUVs to pickups.   12. _____

13. If 400 people were surveyed, how many drove cars?   13. _____

14. If 400 people were surveyed, how many drove pickups?   14. _____

The ages of customers in a store are given in the histogram. Use the histogram to answer questions 15 and 16.

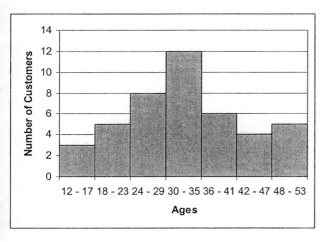

15. How many customers are between 30 and 35 years old?   15. _____

16. How many customers are 23 years old or less?   16. _____

Chapter 8 Practice Test A

17. The Life Expectancies of 22 countries are listed below. Use the ages to complete the frequency distribution table.

83	81	80	78	77
76	75	74	74	73
77	78	79	80	80
80	78	82	82	82
81	78			

Class Intervals (Life Expectancies)	Tally	Class Frequencies (Number of Countries)
73 - 74		
75 - 76		
77 - 78		
79 - 80		
81 - 82		
83 - 84		

18. Use the results of Exercise 17 to draw a histogram.

Find the coordinates of each point in the graph below.

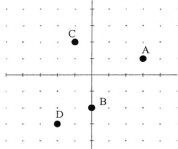

19. A   19. _____

20. B   20. _____

21. C   21. _____

22. D   22. _____

Chapter 8 Practice Test A

Complete and graph the ordered-pair solutions of each given equation.

23. $y = -x$
$(0, \ ), (1, \ ), (\ , 3)$

24. $y = 2x - 3$
$(1, \ ), (2, \ ), (0, \ )$

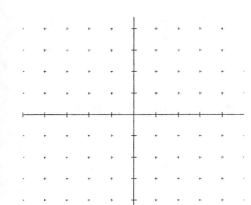
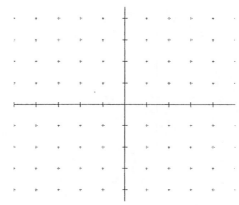

Graph each linear equation.

25. $x + y = -2$

26. $y = 2$

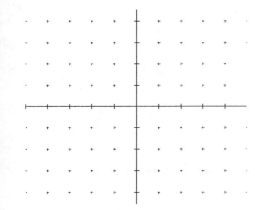
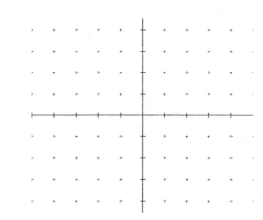

27. $x = -3$

28. $y = x - 1$

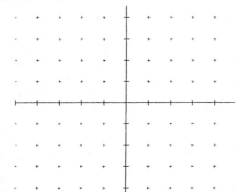
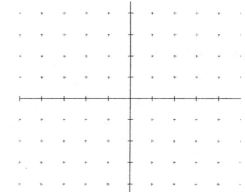

Chapter 8 Practice Test A

29. $y = \dfrac{1}{2}x + 1$ **30.** $2x - y = 3$

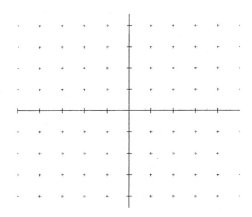   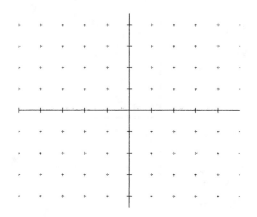

31. Draw a tree diagram for flipping a coin twice.

A bag contains 4 white, 3 yellow and 5 green marbles. If a marble is drawn out, find the following probabilities.

32. getting a white marble   **32.** _____

33. getting a yellow marble   **33.** _____

Chapter 8 Practice Test B

The following pictograph shows the number of goldfish one pet store sold the first three months of 2006. Use this graph to answer questions 1 - 3.

Month	Goldfish Sold
January	
February	
March	

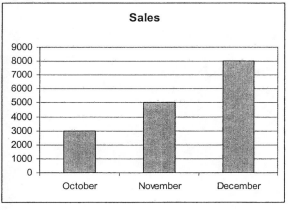

= 100 goldfish

1. In what month were the most goldfish sold?
    a. January   **b.** February   **c.** March

2. How many goldfish were sold in February?
    a. 5   **b.** 700   **c.** 500   **d.** 600

3. How many more goldfish were sold in January then in March?
    a. 300   **b.** 3   **c.** 200   **d.** 2

The bar graph shows the sales for a toy store for the last three months of 2006. Use the bar graph to answer questions 4 - 7.

Sales

9000		
8000		
7000		
6000		
5000		
4000		
3000		
2000		
1000		
0		
October	November	December

4. What were the sales of November?
    a. $4000   **b.** $3000   **c.** $5000   **d.** $8000

5. What month had the most sales?
    a. December   **b.** November   **c.** October

6. Which month had the fewest sales?
    a. December   **b.** November   **c.** October

Chapter 8 Practice Test B

7. How much more were the sales in December than in November?
   a. $6000   **b.** $8000   **c.** $3000   **d.** $5000

The following line graph shows the temperature for Lubbock, Texas, October 31, 2006.
Use the graph to answer question 8 - 10.

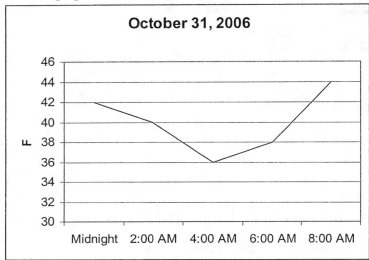

8. What was the temperature at 4:00 AM?
   a. 40°   **b.** 42°   **c.** 36°   **d.** 38°

9. At what time did the highest temperature occur?
   a. 8:00 AM   **b.** Midnight   **c.** 4:00 AM   **d.** 2:00 AM

10. What was the change in temperature between midnight and 4:00 AM?
   a. 8°   **b.** 4°   **c.** 36°   **d.** 6°

Chapter 8 Practice Test B

The circle graph shows the result of a survey of students in a statistics class. Use the graph to answer questions 11 - 14.

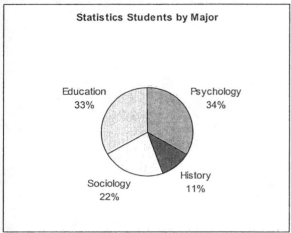

11. What major had the most students?
    a. Education   **b.** Psychology   **c.** History   **d.** Education

12. What major had the least number of students?
    a. Education   **b.** Psychology   **c.** History   **d.** Education

13. If there were 200 students, who many were psychology majors?
    a. 68   **b.** 34   **c.** 66   **d.** 22

14. If there were 200 students, how many were education majors?
    a. 68   **b.** 34   **c.** 66   **d.** 22

Chapter 8 Practice Test B

The grades on the first test in college algebra are given in the histogram. Use the histogram to answer questions 15 - 18.

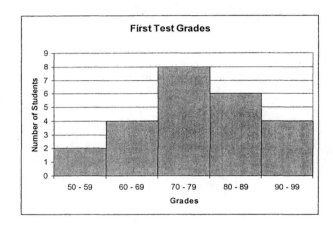

15. How many students made between 60 and 69?
   a. 4   **b.** 8   **c.** 6   **d.** 2

16. How many students made an A (between 90 and 99)?
   a. 4   **b.** 8   **c.** 6   **d.** 2

17. How many students made 70 or above?
   a. 4   **b.** 8   **c.** 18   **d.** 24

18. How many students made 69 or below?
   a. 4   **b.** 8   **c.** 6   **d.** 2

Chapter 8 Practice Test B

Find the coordinates of each point on the graph below.

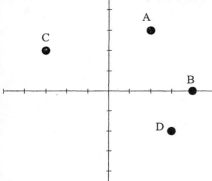

19. A
 a. (3, 2)   b. (2, 3)   c. (2, 0)   d. (2, 4)

20. B
 a. (4, 1)   b. (1, 4)   c. (4, 0)   d. (0, 4)

21. C
 a. (3, 2)   b. (−3, −2)   c. (3, −2)   d. (−3, 2)

22. D
 a. (3, 2)   b. (−3, −2)   c. (3, −2)   d. (−3, 2)

Use the equation $y = x + 3$ for 23 - 25. Complete the ordered pair solution.

23. (0,   )
 a. (0, 3)   b. (0, −3)   c. (0, 0)   d. (0, 1)

24. (1,   )
 a. (1, 1)   b. (1, −3)   c. (1, 4)   d. (1, 2)

25. (, 0)
 a. (3, 0)   b. (−3, 0)   c. (1, 0)   d. (−1,)

Chapter 8 Practice Test B

26. Use the points from problems 23 - 25 to plot the line for the equation $y = x + 3$.

a.

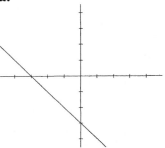

b.

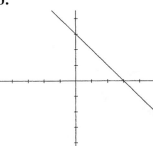

c.

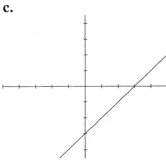

d.

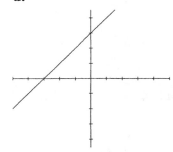

Match the graph to the given equation.

27. $x = 4$

a.

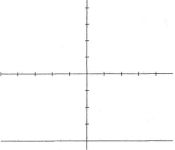

b.

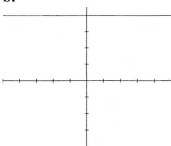

c.

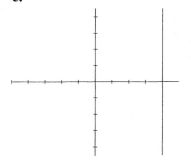

d.

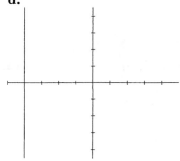

Chapter 8 Practice Test B

28. $y = -4$

a.

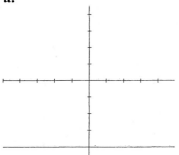

b.

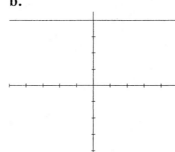

c.

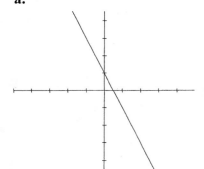

d.

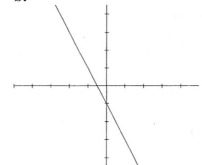

29. $y = 2x - 1$

a.

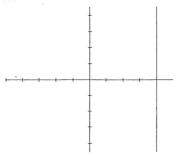

b.

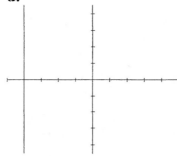

c.

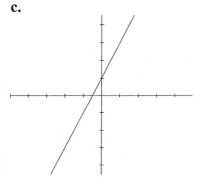

d.

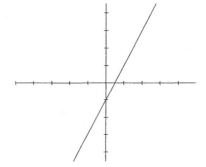

Chapter 8 Practice Test B

30. $x + 2y = 4$

   a.

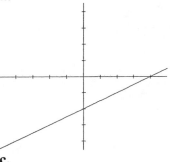

   b.

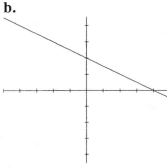

   c.

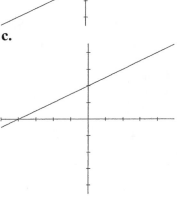

   d.

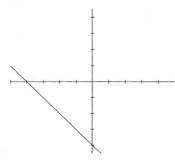

One ball is drawn from a bag containing six balls that have been numbered 1 - 6. Find each of the following probabilities.

31. Probability of getting a 5.

   a. $\dfrac{1}{5}$ **b.** $\dfrac{1}{2}$ **c.** $\dfrac{1}{6}$ **d.** 1

32. getting an even number

   a. $\dfrac{1}{5}$ **b.** $\dfrac{1}{2}$ **c.** $\dfrac{1}{6}$ **d.** 1

33. getting a number less than 3

   a. $\dfrac{1}{3}$ **b.** $\dfrac{1}{2}$ **c.** $\dfrac{1}{6}$ **d.** 1

9.1 Geometry and Measurement

Learning Objectives
    A. Identify lines, line segments, rays, and angles.
    B. Classify angles as acute, right, obtuse, or straight.
    C. Identify complementary and supplementary angles.
    D. Find measures of angles.

Vocabulary
line, line segment, ray, angle, vertex

Objective A Identify lines, line segments, rays, and angles.

Identify each figure as a line, a ray, a line segment, or an angle. Name the figure.

Example 1
a)

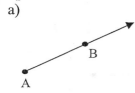

b)

c)

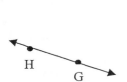

d)

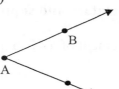

Example 2 List other ways to name angle *x*.

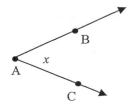

1a) $\overrightarrow{AB}$ is a ray, b) $\overline{ZY}$ is a line segment, c) $\overleftrightarrow{HG}$ line, d) $\angle BAC$ is an angle, 2) $\angle BAC$, $\angle CAB$,
$\angle A$

Examples 9.1 (cont'd)

Objective B Classify angles as acute, right, obtuse, or straight.

Example 3 Classify each angle as acute, right, obtuse, or straight.

a)   b)

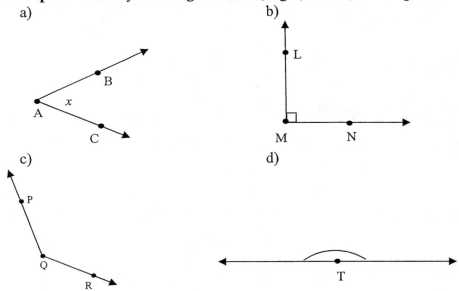

c)   d)

Objective C Identify complementary and supplementary angles.

Example 4 Find the complement of a 47° angle.

Example 5 Find the supplement of a 132° angle.

Objective D Find measures of angles.

Example 6 Find the measure of $\angle x$ then classify $\angle x$ as an acute, obtuse, or right angle. $\angle PQR = 135°$ and $\angle SQR = 48°$.

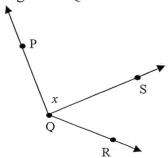

3a) acute, b) right, c) obtuse, d) straight, 4) 43°, 5) 48°, 6) 87°, acute

Examples 9.1 (cont'd)

Example 7 Find the measures of ∠a, ∠b, and ∠c if ∠d is 43°.

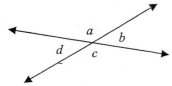

Example 8 Given that l ∥ m and that the measure of ∠h is 132°, find the measures of ∠e, ∠f, and ∠g.

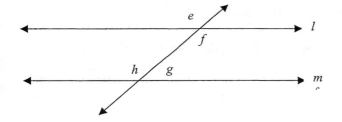

7) ∠a = 137°, ∠b = 43°, ∠c = 137°, 8) ∠e = 132°, ∠f = 132°, ∠g = 48°

Practice Set 9.1

Use the choices below to fill in each blank.

acute	adjacent	angle	
line	obtuse	ray	vertex

1. Two angles that share a common side are called _____ angles.

2. A(n) _____ is made up two rays that share a common endpoint and the common endpoint is called the _____.

3. A(n) _____ is part of a line with one endpoint.

4. A(n) _____ is a set of points extending indefinitely in two directions.

5. A(n) _____ angle measures between 90° and 180°.

6. A(n) _____ angle measures between 0° and 90°.

Objective A Identify lines, line segments, rays, and angles.

Identify each figure as a line, a ray, a line segment, or an angle. The name the figure using the given points

7.

7. _____

8.

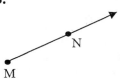

8. _____

9.

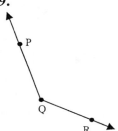

9. _____

Exercise Set 9.1 (cont'd)

10.

10. _____

11. List two other ways to name ∠*x*.

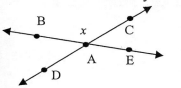

11. _____

Objective B Classify angles as acute, right, obtuse, or straight.

12.

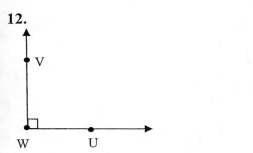

12. _____

13.

T

13. _____

14.

G

x

H

I

14. _____

15.

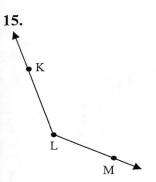

K

L

M

15. _____

Exercise Set 9.1 (cont'd)

Objective C Identify complementary and supplementary angles.

16. Find the complement of 23°.   16.  _____

17. Find the supplement of 57°.   17.  _____

Objective D Find measures of angles.

Find the measure of angle *y* in each figure.

18. ∠ *x* is 47°   18. _____

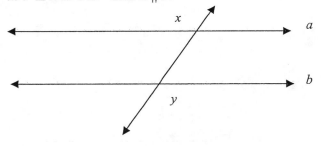

19. ∠ *x* is 119° and *a* ‖ *b*.   19. _____

Extensions

20. Find the measures of angle *j* through *m* given the   20. _____
lines *r* and *s* are parallel.

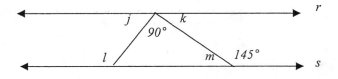

9.2 Perimeter

Learning Objectives
A. Use formulas to find perimeters.
B. Use formulas to find circumference.

Objective A Use formulas to find perimeters.

Example 1 Find the perimeter of the rectangle below.

6 inches

14 inches

Example 2 Find the perimeter of a rectangle with a length of 10 inches and a width of 5 inches.

Example 3 Find the perimeter of a square field that is 80 yards on each side.

Example 4 Find the perimeter of a triangle that has sides of 4 inches, 6 inches, and 10 inches.

Example 5 Find the perimeter of the trapezoid given below.

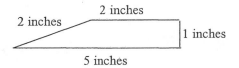

2 inches

2 inches

1 inches

5 inches

1) 40 inches, 2) 30 inches, 3) 320 yards, 4) 20 inches, 5) 10 inches

Examples 9.2 (cont'd)

Example 6 Find the perimeter of room shown below.

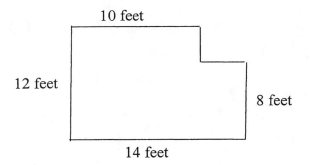

Example 7 Calculate the cost to put a border around a rectangular garden that is 4 yards by 5 yards if the border costs $3.60 a yard.

Objective B Use formulas to find circumferences.

Example 8 Find the circumference of a circle with a diameter of 12 feet. Find the exact circumference than use 3.14 to approximate the circumference.

6) 52 feet, 7) $64.80, 8) 12 π feet, 37.68 feet

Name: Date:
Instructor: Section:

Practice Set 9.2

Use the choices below to fill in each blank.

circumference   **diameter**   **perimeter**   **pi**

1. The distance around a circle is called the _____.

2. The sum of the lengths of the sides of a polygon is called the _____.

3. If you double the radius of a circle, you get the circle's _____.

4. The exact ratio of circumference to diameter is _____.

Objective A Use formulas to find perimeters.

Find the perimeter of each figure.

5. Rectangle   5. _____

6. Parallelogram   6. _____

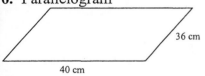

7. A parallelogram with height of 7 yards and length   7. _____
   of 4 yards.

8. A triangle with sides of 8 in, 10 in, and 14 in.   8. _____

9.   9. _____

10.   10. _____

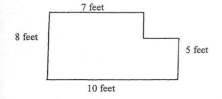

Practice Set 9.2 (cont'd)

11. Regular hexagon

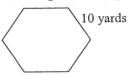

10 yards

11. _____

12. Regular pentagon.

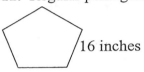

16 inches

12. _____

13. A rectangle with length of 8 in and width of 5 in.

13. _____

14. A square with sides of 8 miles.

14. _____

15.

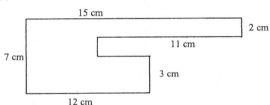

15. _____

Objective B Use formulas to find circumferences.

Find the circumference of each circle. Give the exact circumference and then use $\pi \approx 3.14$ to approximate the circumference.

16. Diameter = 8 inches

16. _____

17. Radius = 9 inches

17. _____

18. Diameter = 3 feet

18. _____

19. Radius = 3.5 in

19. _____

Extensions

20. Find the distance around the running track shown below. Use 3.14 to approximate pi.

20. _____

15 feet

25 feet

9.3 Area, Volume, and Surface area

Learning Objectives
    A. Find the area of plane figures.
    B. Find the volume and surface area of solids.

Objective A Find the area of plane figures.

    Example 1 Find the area of a triangle with height of 6 inches and base of 14 inches.

    Example 2 Find the area of a parallelogram with height of 17 inches and base of 35 inches.

    Example 3 Find the area of the figure.

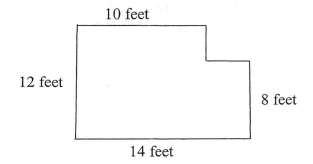

    Example 4 Find the area of a circle with a radius of 6 feet. Find the exact area and an approximation by using 3.14 for π.

1) 42 square inches, 2) 595 square inches, 3) 152 square feet, 4) 36π or 113.04 square feet

Examples 9.3 (cont'd)

Objective B Find volume and surface area of solids.

Example 5 Find the volume and surface area of a rectangular box that is 16 inches long, 8 inches wide, and 4 inches tall.

Example 6 Find the volume and the surface area of a ball of radius 4 inches. Give the exact volume and surface area and then use the approximation $\dfrac{22}{7}$ for π .

Example 7 Find the volume of a can that has a 2 inch radius and a height of 5 inches. Give the exact volume and then use the approximation $\dfrac{22}{7}$ for π .

Example 8 Find the approximate volume of a cone that has a height of 15 cm and a radius of 4 cm. Use 3.14 for an approximation of π .

5) 512 cubic inches, 448 square inches, 6) $\dfrac{256}{3}\pi$ cubic inches or 268.19 cubic inches, 64π

or $201\dfrac{1}{7}$ sq in., 7) 20π or $62\dfrac{6}{7}$ cu in. 8) 251.2 cu cm

Name:

Instructor:

Date:

Section:

Practice Set 9.3

Use the choices below to fill in each blank.

cubic	square	surface area	volume

1. Surface area is measured in _____ units.

2. Volume is measured in _____ units.

3. Area is measured in _____ units.

4. The sum of the areas of the faces of a polyhedron is the _____.

5. The measure of the space of a solid is its _____.

Objective A Find the area of plane regions.

Find the area of each figure.

6. a rectangle with length of 8 m and width of 6 m   6. _____

7. a triangle with height of 8 in and base of 7 in.   7. _____

8. parallelogram with height of 9 in and base of 10 in.   8. _____

9. circle with radius of 8 ft, use $\dfrac{22}{7}$ for π.   9. _____

10. trapezoid with bases of 10 and 14 in, and height of   10. _____

    7 inches

11. trapezoid with bases of 8 and 11 ft and height of   11. _____

    12 ft

12. parallelogram with height of 6 cm and base of 15 cm   12. _____

Practice Set 9.3 (cont'd)

13.

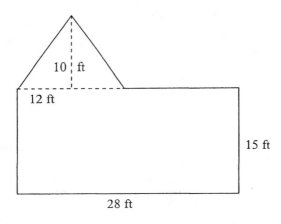

Objective B Find the volume and surface area of solids.

Find the volume and surface area of each solid. For formulas containing π, use $\dfrac{22}{7}$.

14. a box with sides of 6 and 8 in, and height of 3 in

14. _____

15. a cube with sides of 9 cm

15. _____

16. a cone with radius of 3 yd and height of 4 yd

16. _____

17. a ball with diameter of 10 inches

17. _____

18. volume only of a cylinder with radius of 4 in and height of 5 in

18. _____

19. volume of a square based pyramid with height of 9 feet and each side of the base 5 feet

19. _____

Extensions

20. Find the area of the figure below. Use 3.14 for pi. Round to whole number.

20. _____

15 feet

25 feet

9.4 Linear Measurement

Learning Objectives
    A. Define U.S. units of length and convert from one unit to another.
    B. Use mixed units of length.
    C. Perform arithmetic operations on U.S. units of length.
    D. Define metric units of length and convert from one unit to another.
    E. Perform arithmetic operations on metric units of length.

Objective A Define U.S. units of length and convert from one unit to another.

    Example 1 Convert 12 feet to inches.

    Example 2 Convert 10 feet to yards.

    Example 3 A ring-necked pigeon has a 15-inch wing spread. Convert 15 inches to feet. Use decimals in your final answer.

Objective B Use mixed units of length.

    Example 4 Convert: 140 in. = _____ ft _____ in.

    Example 5 Convert 6 feet 4 inches to inches.

Objective C Perform arithmetic operations on U.S. units of length.

    Example 6 Add 6 ft 2 in. and 4 ft and 10 in.

1) 144 in., 2) $3\frac{1}{3}$ yd, 3) 1.25 ft, 4) 11 ft 8 in., 5) 76 in., 6) 11 ft

Examples 9.4 (cont'd)

Example 7 Multiply 3 ft 4 in. by 4.

Example 8 A rope of length 5 yd 1 ft has a length of 1 yd and 2 ft cut from one end. Find the length of the remaining rope.

Objective D Define metric units of length and convert from one unit to another.

Example 9 Convert 4.6 m to cm.

Example 10 Convert 125,000 mm to meters.

Objective E Perform arithmetic opertations on metric system units of length.

Example 11 Subtract 560 m from 1.4 km.

Example 12 Multiply 2.6 mm by 5.

Example 13 Jorge Lopez is 0.6 meters tall six months ago. Since then he has grown 20 cm. Find his current height in meters.

7) 13 ft 4 in., 8) 3 yd 2 ft, 9) 460 cm, 10) 125 m, 11) 0.84 km or 840 m, 12) 13 mm, 13) 0.8 m

Practice Set 9.4

Objective A Define U.S. units of length and convert form one unit to another..

Convert each unit as indicated.

1. 72 in. to feet

1. _____

2. 8 yd to feet

2. _____

3. 26,400 ft to miles

3. _____

4. $6\frac{1}{2}$ ft to inches

4. _____

Objective B Use mixed rates of length.

5. 20 ft

5. _____ yd _____ ft

6. 65 in

6. _____ ft _____ in.

7. 5 yd 2 ft to ft

7. _____

8. 10,860 ft to miles

8. _____ mi _____ ft

Objective C Perform arithmetic operations on U.S. units of length.

Perform each indicated operation.

9. 4 ft 8 in. + 3 ft 6 in.

9. _____

10. 3 yd 2 ft + 4 yd 2 ft

10. _____

11. 16 ft 5 in − 9 ft 7 in

11. _____

Practice Set 9.4 (cont'd)

12. 15 yd 2 ft × 4   12. _____

Objective D Define metric units of length and convert from one unit to another.

Convert as indicated.

13. 48 m to centimeters   13. _____

14. 1200 mm to meters   14. _____

15. 5640 cm to meters   15. _____

16. 8 km to meters   16. _____

17. 0.008 m to millimeters   17. _____

Objective E Perform arithmetic operations on metric units of length.

18. 70 cm + 8.9 m (answer in cm)   18. _____

19. 8.4 km ÷ 4   19. _____

Extensions

20. Find the area of the following rectangle. Give answer **20.** _____
in square feet.

2 yd

2 ft

9.5 Weight and Mass

Learning Objectives
  A. Define U.S. units of weight and convert from one unit to another.
  B. Perform arithmetic operations on U.S. units of weight.
  C. Define metric units of mass and convert from one unit to another.
  D. Perform arithmetic operations on metric units of mass.

Vocabulary
weight, mass

Objective A Define U.S. units of weight and convert from one unit to another.

  Example 1 Convert 8000 pounds to tons.

  Example 2 Convert 4 pounds to ounces.

  Example 3 Convert: 28 ounces = _____ lb _____ oz.

Objective B Perform arithmetic operations on U.S. units of weight.

  Example 4 Subtract 1 ton 1400 lb from 4 tons 800 lb.

  Example 5 Divide 7 lb 4 oz by 2.

1) 4 tons, 2) 64 oz, 3) 1 lb 12 oz, 4) 2 tons 1400 lb, 5) 3 lb 10 oz

Examples 9.5 (cont'd)

Example 6 Julie weighed 6 pounds 10 oz at birth. By her first birthday, she had gained 10 lb 12 oz. Find her weight at age 1 year.

Objective C Define metric units of mass and convert from one unit to another.

Example 7 Convert 1.6 kg to grams.

Example 8 Convert 1.75 cg to grams.

Objective D Perform arithmetic operations on metric units of mass

Example 9 Subtract 2.6 dg from 1.1 g.

Example 10 An elevator has a weight limit of 1000 kg. A sign posted indicates that the maximum capacity of the elevator is 14 persons. What is the average allowable weight for each passenger, rounded to the nearest kilogram?

6) 17 lb 6 oz, 7) 1600 g, 8) 0.0175 g, 9) 0.84 g or 8.4 dg, 10) 71 kg

Practice Set 9.5

Use the choices below to fill in each blank.

mass   **weight**

1. The measure of the amount of substance in an object is called _____.

2. The measure of the pull of gravity is called _____.

Fill in the correct number.

4. One pound equals _____ ounces.

5. One ton equals _____ pounds.

Objective A Define U.S. units of weight and convert from one unit to another.

Convert as indicated. Give answer as a fraction if needed.

5. 3 pounds to ounces   5. _____

6. 14,000 pounds to tons   6. _____

7. 4500 pounds to tons   7. _____

8. 1.5 tons to pounds   8. _____

9. $3\frac{1}{2}$ pounds to ounces   9. _____

Objective B Perform arithmetic operations on units of weight.

Perform each indicated operation.

10. 7 lb 10 oz + 5 lb 6 oz   10. _____

11. 8 lb 4 oz − 5 lb 8 oz   11. _____

Practice Set 9.5 (cont'd)

12. 3 lb 6 oz × 4

12. _____

13. 5 tons 1200 lb ÷ 4

13. _____

Objective C Define metric units of mass and convert from one unit to another.

Convert as indicated.

14. 650 g to kilograms

14. _____

15. 50 kg to grams

15. _____

16. 3.35 kg to grams

16. _____

17. 7.21 g to milligrams

17. _____

Objective D Perform arithmetic operations on units of mass.

Perform the indicated operation.

18. 3.2 g + 4.9 g

18. _____

19. 3 g − 1720 mg

19. _____

20. 1.5 kg × 6

20. _____

9.6 Capacity

Objective A Define U.S. units of capacity and convert from one unit to another.

   Example 1 Convert 6 quarts to gallons.

   Example 2 Convert 8 cups to quarts.

Objective B Perform arithmetic operations on U.S. units of capacity.

   Example 3 Subtract 3 qt from 2 gal 1 qt.

   Example 4 An aquarium contains 3 gal 1 qt of water. If 1 gal 3 qt of water is added, what is the total amount of water in the aquarium?

Objective C Define metric units of capacity and convert from one unit to another.

   Example 5 Convert 1400 ml to liters.

1) $1\frac{1}{2}$ gal, 2) 2 qt, 3) 1 gal 2 qt, 4) 5 gal, 5) 1.4 L

Examples 9.6 (cont'd)

Example 6 Convert 0.15 dl to milliliters.

Objective D Perform arithmetic operations on metric system units of capacity.

Example 7 Add 1600 ml to 1.4 L.

Example 8 A patient hooked up to an IV in the hospital is to receive 8.5 ml of medication every hour. How much medication does the patient receive in 4 hours?

6) 15 ml, 7) 3 L or 3000 ml, 8) 34 ml

Practice Set 9.6

Use the choices below to fill in each blank.

capacity   **liter**

1. _____ units are usually used to measure liquids.

2. The basic unit of capacity in the metric system is the _____.

Objective A Define U.S. units of capacity and convert from one unit to another.

Convert. Give answers in fractions when necessary.

3. 16 fluid ounces to cups   3. _____

4. 7 cups to pints   4. _____

5. 12 quarts to gallons   5. _____

6. 36 cups to quarts   6. _____

7. 50 qt =   7. _____ gal _____ qt

Objective B Perform arithmetic operations on U.S. units of capacity.

Perform the indicated operation.

8. 3 gal 3 qt + 7 gal 2 qt   8. _____ gal _____ qt

9. 2 c 5 fl oz + 3 c 6 fl oz   9. _____ c _____ oz

10. 3 gal 1 qt − 2 qt 1 pt   10. ____gal____ gt ____ pt

11. 3 gal 2 qt × 3   11. _____

Practice Set 9.6 (cont'd)

Objective C Define metric units of capacity and convert from one unit to another.

Convert.

12. 5 L to milliliters

12. _____

13. 9000 ml to liters

13. _____

14. 1500 L to kiloliters

14. _____

15. 0.14 kl to liters

15. _____

16. 0.98 kl to liters

16. _____

Objective D Perform arithmetic operations on metric units of capacity.

Perform the indicated operation.

17. 31 L + 17.9 L

17. _____

18. 4.4 L – 350 ml

18. _____

19. 750 ml × 4

19. _____

20. 6.4 L ÷ 0.5

20. _____

9.7 Temperature and Conversions Between the U.S. and Metric Systems

---

Learning Objectives
    A. Convert between the U.S. and metric systems.
    B. Convert temperatures from degrees Celsius to degrees Fahrenheit.
    C. Convert temperatures from degrees Fahrenheit to degrees Celsius.

Objective A Convert between the U.S. and metric systems.

    Example 1 Convert 18 cm to in.

    Example 2 Convert 4.2 pounds to kg. Round to two decimal places.

    Example 3 Convert 8 fl oz to ml. Round to two decimal places.

Objective B Covert temperatures from degrees Celsius to degrees Fahrenheit.

    Example 4 Convert 20° C to degrees Fahrenheit

    Example 5 Convert 32°C to degrees Fahrenheit. Round to one decimal place.

1) 7.09 in., 2) 1.91 kg,, 3) 236.56 ml, 4) 68°F, 5) 89.6°F

Examples 9.7 (cont'd)

Objective C Convert temperatures from degrees Fahrenheit to degrees Celsius.

Example 7 Convert 41° F to degrees Celsius.

Example 8 Convert 120° F to degrees Celsius. Round to one decimal place.

Example 9 Convert 101.4°F to degrees Celsius. Round to one decimal place.

6) 5°C, 7) 48.9°C, 8) 38.6°C

Practice Set 9.7

Objective A Convert between U.S. and metric systems.

Convert. Round to two decimal places if necessary.

1. 800 ml to fl oz   1. _____

2. 16 liters to quarts   2. _____

3. 90 inches to centimeters   3. _____

4. 95 miles to kilometers   4. _____

5. 500 grams to ounces   5. _____

6. 200 kilograms to pounds   6. _____

7. 108 kilometers to miles   7. _____

8. 6.2 meters to feet   8. _____

9. 12.5 liters to gallons   9. _____

10. 200 milliliters to fluid ounces   10. _____

Practice Set 9.7 (cont'd)

11. 60 pounds to kilograms

11. _____

12. 20 ounces to grams

12. _____

Objective B Convert temperatures from degrees Celsius to degrees Fahrenheit.

Convert. Round to one decimal place if necessary.

13. 5°C to degrees Fahrenheit

13. _____

14. 27°C to degrees Fahrenheit

14. _____

15. 88°C to degrees Fahrenheit

15. _____

16. 50° C to degrees Fahrenheit

16. _____

Objective B Convert temperatures from degrees Fahrenheit to degrees Celsius.

Convert. Round to one decimal place if necessary.

17. 40°F to degrees Celsius

17. _____

18. 212°F to degrees Celsius

18. _____

19. 56°F to degrees Celsius

19. _____

20. 99°F to degrees Celsius

20. _____

Chapter 9 Vocabulary Reference Sheet

Term	Definition	Example
Section 9.1		
Line	A set of points extending indefinitely in two directions.	H, G is a line $\left(\overleftrightarrow{HG}\right)$.
Line segment	A piece of a line with 2 endpoints.	Z, Y is a line segment $\left(\overline{ZY}\right)$.
Ray	A part of a line with one endpoint. A ray extends indefinitely in one direction.	A, B is a ray $\left(\overrightarrow{AB}\right)$.
Angle	Two rays that share a common endpoint.	A, B, C is angle ($\angle$BAC).
Vertex	The common endpoint of two rays that form an angle.	A, B, C A is the vertex of $\angle$BAC.
Section 9.5		
Weight	A measure of the pull of gravity.	Pounds are one unit used to measure weight.
Mass	A measure of the substance of an object.	Grams are one unit used to measure mass.

Chapter 9 Practice Test A

1. Find the complement of a 18° angle.   1.

2. Find the supplement of a 21° angle.   2. _____

3. Find the measure of angle x if $\angle y$ is 29 ° and $\angle$ ABC is 59°.   3. _____

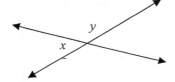

4. Find the measure of angle x if $\angle y$ is 132°.   4. _____

5. Find the measure of angle x if $\angle y$ is 125°.   5. _____

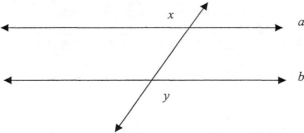

6. Find the diameter of a sphere if the radius is 6 inches.   6. _____

7. Find the radius of a circle if the diameter is 24 ft.   7. _____

Find the perimeter (or circumference) and area of each figure. Use 3.14 as an approximation for pi.

8. a circle with radius 12 in.   8. _____

9. a rectangle with length of 1.5 ft and width of 0.98 ft.   9. _____

Chapter 9 Practice Test A (cont'd)

10.

20 feet

22 feet

16 feet

28 feet

10. _____

Find the volume of each solid.

11. A cylinder with radius 2 in and height 6 in. Use $\frac{22}{7}$ for pi.

11. _____

12. A box with length of 8 in., width of 6 in. and height of 4 in.

12. _____

13. Find the perimeter of a square with sides of 12 in.

13. _____

14. How much dirt is needed to fill a rectangular hole that is 6 feet by 5 feet by 7 feet.

14. _____

15. How much border is needed to go around a garden that is 20 ft by 12 ft? If the border costs $3.50 a foot, how much will it cost for the border?

15. _____

Convert. Write answers in the U.S. system as a fraction if necessary.

16. 68 in

16. _____ ft _____ in

17. $1\frac{1}{2}$ gallons to quarts

17. _____

18. 80 oz to pounds

18. _____

Chapter 9 Practice Test A (cont'd)

19. 3.2 tons to pounds

19. _____

20. 30 pt to gallons

20. _____

21. 30 mg to grams

21. _____

22. 1.4 kg to grams

22. _____

23. 2.1 cm to millimeters

23. _____

24. 8 dg to grams

24. _____

25. 0.93 L to milliliters

25. _____

Perform each indicated operation.

26. 4 qt 1 pt + 3 qt 1 pt

26. _____

27. 6 lb 2 oz − 3 lb 8 oz

27. _____

28. 2 ft 9 in × 4

28. _____

29. 2 gal 2 qt ÷ 2

29. _____

30. 1.8 km + 298 m

30. _____

31. 10 cm − 96 mm

31. _____

32. Convert 75°F to degrees Celsius. Round to one decimal place.

32. _____

33. Convert 18.6°C to degrees Fahrenheit. Round to one decimal place.

33. _____

Chapter 9 Practice Test B

1. Find the complement of a 25° angle.
  a. 75°   **b.** 65°   **c.** 155°   **d.** 115°

2. Find the supplement of a 67° angle.
  a. 155°   **b.** 23°   **c.** 113°   **d.** 33°

3. Find the measure of angle x if $\angle y$ is 67 ° and $\angle$ ABC is 122°.

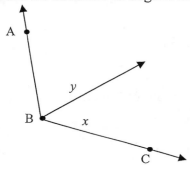

  a. 58°   **b.** 61°   **c.** 151°   **d.** 55°

4. Find the measure of angle x if $\angle y$ is 145°.

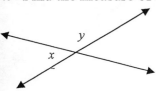

  a. 35°   **b.** 61°   **c.** 55°   **d.** 65°

5. Find the measure of angle x if $\angle y$ is 132°.

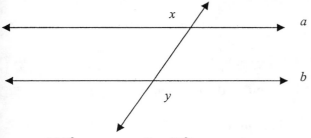

  a. 132°   **b.** 48°   **c.** 42°   **d.** 32°

Chapter 9 Practice Test B (cont'd)

6. Find the circumference of a circle with radius of 14 in. Use $\frac{22}{7}$ for π.

   a. 44 in.   **b.** 22 in.   **c.** 88 in.   **d.** 28 in.

7. Find the area of a circle with radius of 14 in. Use $\frac{22}{7}$ for π.

   a. 154 sq in.   **b.** 2462 sq in.   **c.** 192 sq in.   **d.** 616 sq in.

8. Find the area of a rectangle with length of 12 feet and width of 5 feet.
   a. 17 sq ft   **b.** 60 sq ft   **c.** 30 sq ft   **d.** 34 sq ft

9. Find the perimeter of a rectangle with length of 12 feet and width of 5 feet.
   a. 17 ft   **b.** 60 ft   **c.** 30 ft   **d.** 34 ft

10. Find the area of the figure below.

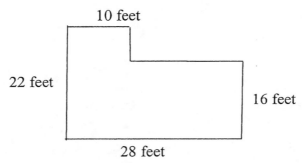

   a. 636 sq ft   **b.** 508 sq ft   **c.** 676 sq ft   **d.** 100 sq ft

Find the volume of each solid.

11. A cylinder with radius 4 in and height 6 in. Use $\frac{22}{7}$ for π.

   a. 452.6 cu in.   **b.** 75.9 cu in.   **c.** 301.7 cu in. **d.** 1762 cu in.

12. A box with length of 10 in., width of 12 in. and height of 6 in.
   a. 353 cu ft   **b.** 504 cu ft   **c.** 240 cu ft   **d.** 720 cu ft

Chapter 9 Practice Test B

13. Find the perimeter of a square with sides of 12 in.
    a. 24 in   **b.** 48 in.   **c.** 12 in.   **d.** 144 in.

14. How much dirt is needed to fill a rectangular hole that is 8 feet by 4 feet by 6 feet.
    a. 104 cu ft   **b.** 208 cu ft   **c.** 384 cu ft   **d.** 192 cu ft

15. A border is being built to go around a garden that is 10 ft by 15 ft. If the border costs $3.50 a foot, how much will it cost for the border?
    a. $525   **b.** $87.50   **c.** $175   **d.** $270

Convert.

16. 75 in
    a. 6 ft 4 in   **b.** 6 ft 3 in   **c.** 5 ft 15 in   **d.** 6 ft 6 in

17. $3\frac{1}{2}$ gallons to quarts
    a. 14 qt   **b.** 7 qt   **c.** 28 qt   **d.** 10 qt

18. 96 oz to pounds
    a. 1536 lb   **b.** 6 lb   **c.** 12 lb   **d.** 8 lb

19. 1.8 tons to pounds
    a. 28.8 lb   **b.** 1800 lb   **c.** 3600 lb   **d.** 36 lb

20. 16 pt to gallons
    a. 1 gal   **b.** 4 gal   **c.** 8 gal   **d.** 2 gal

21. 65 mg to grams
    a. 0.065 g   **b.** 65,000 g   **c.** 0.65 g   **d.** 650 g

22. 2.4 kg to grams
    a. 240 g   **b.** 2400 g   **c.** 0.0024 g   **d.** 0.024 g

Chapter 9 Practice Test B

23. 4.5 cm to millimeters
    a. 450 mm   **b.** 4500 mm   **c.** 0.45 mm   **d.** 45mm

24. 9 dg to grams
    a. 9000 g   **b.** 90 g   **c.** 0.9 g   **d.** 0.009 g

25. 1.2 L to milliliters
    a. 120 ml   **b.** 0.12 ml   **c.** 1200 ml   **d.** 0.0012 ml

Perform each indicated operation.

26. 5 qt 1 pt + 3 qt 1 pt
    a. 9 qt   **b.** 8 qt   **c.** 8 qt 2 pt   **d.** 10 qt

27. 5 lb 8 oz – 3 lb 10 oz
    a. 2 lb 2 oz   **b.** 1 lb 8 oz   **c.** 1 lb 6 oz   **d.** 1 lb 14 oz

28. 3 ft 10 in $\times$ 2
    a. 7 ft 10 in.   **b.** 7 ft 8 in.   **c.** 6 ft 10 in.   **d.** 6 ft 20 in.

29. 7 gal 2 qt $\div$ 2
    a. 3 gal 3 qt   **b.** 3 gal 1 qt   **c.** 3 gal 6 qt   **d.** 4 gal 8 qt

30. 3 km +400 m
    a. 3.4 m   **b.** 3.4 km   **c.** 403 m   **d.** 403 km

31. 8 km – 1000 m
    a. 0.7 km   **b.** 7.9 km   **c.** 7 km   **d.** 7.0 m

32. Convert 50°F to degrees Celsius.
    a. 122°C   **b.** 63°C   **c.** 80°C   **d.** 10°C

33. Convert 40°C to degrees Fahrenheit.
    a. 104°F   **b.** 4.4°F   **c.** 98°F   **d.** 64°F

10.1 Exponents and Polynomials

Learning Objectives
    A. Add polynomials.
    B. Subtract polynomials.
    C. Evaluate polynomials at given replacement values.

Vocabulary
polynomial, monomial, binomial, trinomial

Objective A Add polynomials.

    Example 1 Add: $(2x - 3) + (x - 7)$

    Example 2 Add: $(3y^2 - 6y) + (4y^2 - 2y + 5)$

    Example 3 Find the sum of $\left(-x^2 + 3x - 1\right)$ and $\left(3x^2 - 9x - 1.8\right)$.

    Example 4 Find the sum of $\left(-y^2 + 7y - 3.4\right)$ and $\left(-2y^2 - 8y + 4.5\right)$. Use a vertical
format.

Objective B Subtract polynomials.

    Example 5 Simplify: $-\left(3x^2 - 8x - 8\right)$

    Example 6 Subtract: $(3a - 7) - (4a + 2)$

1) $3x - 10$, 2) $7y^2 - 8y + 5$, 3) $2x^2 - 6x - 2.8$, 4) $-3y^2 - y + 1.1$, 5) $-3x^2 + 8x + 8$,
6) $-a - 9$

Examples 10.1 (cont'd)

Example 7 Subtract: $\left(6x^2 - 2x + 1\right) - \left(3x^2 - 8\right)$

Example 8 Subtact $\left(-4a^2 - 2a + b\right)$ from $\left(3a^2 - 2a\right)$.

Example 9 Subtract: $\left(5x^2 - 2x - 7\right)$ from $\left(3x^2 - 2x + 8\right)$. Use a vertical format.

Objective B Evaluate polynomials at given replacement values.

Example 10 Find the value of the polynomial $2t^2 - 4t + 1$ when $t = 2$.

Example 11 An object is dropped from the top of a 600-foot-tall building. Its height at time t seconds is given by the polynomial $-16t^2 + 600$. Find the height of the object when $t = 1$ and $t = 4$ seconds.

7) $3x^2 - 2x + 9$, 8) $7a^2 - b$, 9) $-2x^2 + 15$, 10) 1, 11) *584 feet, 344 feet*

Practice Set 10.1

Use the choices below to fill in each blank.

binomial   **monomial**
trinomial   **terms**

1. The polynomial $3x^2 + 5x - 4$ is called a _____.

2. A polynomial $4a - 5$ is called a _____.

3. A polynomial $3x^2 y^4$ is called a _____.

4. In the polynomial $4a - 5$, $4a$ and -5 are called _____.

Objective A Add polynomials.

Add the polynomials.

5. $(3x - 5) + (-5x - 4)$ 5. _____

6. $(9a - 5) + (7a - 4)$ 6. _____

7. $(4x^2 - 7x + 4) + (3x^2 + 9x - 5)$ 7. _____

8. $(3y^2 - 6) + (4y^2 - 3y + 8)$ 8. _____

9. $(4x^2 - 7x + 3) + (-3x^2 - 5x - 4)$ 9. _____

Objective B Subtract polynomials.

Simplify.

10. $-(3x - 5)$ 10. _____

11. $-(3x^2 - 7x - 9)$ 11. _____

Practice Set 10.1 (cont'd)

Subtract the polynomials.

12. $(9a - 12) - (7a - 15)$

12. _____

13. $(3x^2 - 9x - 10) - (4x^2 - 6x + 2)$

13. _____

14. $(10y^2 - 7y + 2) - (3y^2 - 7y + 5)$

14. _____

15. Subtract $(5a^2 - 2a - 6)$ from $(3a^2 - 7a + 12)$

15. _____

Objective C Evaluate polynomials at given replacement values.

Find the value of each polynomial for the given value.

16. $-4x + 1$ for $x = 3$

16. _____

17. $3x^2 - 3x + 5$ for $x = 3$

17. _____

18. $-x^2 - 3x$ for $x = 5$

18. _____

19. $\dfrac{3x^2}{4} - 8$ for $x = 4$

19. _____

Extensions

20. Find the perimeter of the triangle.

20. _____

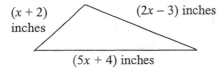

$(x + 2)$ inches $(2x - 3)$ inches

$(5x + 4)$ inches

10.2 Multiplication Properties of Exponents

Learning Objectives
A. Use the product rule for exponents.
B. Use the power rule for exponents.
C. Use the power of a product rule for exponents.

Objective A Use the product rule for exponents.

Example 1 Multiply: $x^3 \cdot x^5$

Example 2 Multiply: $2x^3 \cdot 3x^4$

Example 3 Multiply: $\left(-3a^4b^2\right)\left(16a^2b\right)$

Example 4 Multiply: $3x^2 \cdot 2x \cdot 4x^5$

Objective B Use the power rule for exponents.

Example 5 Simplify: $\left(x^3\right)^4$

Example 6 Simplify: $\left(a^2\right)^4 \cdot \left(a^3\right)^5$

1) x^8, 2) $6x^7$, 3) $-48a^6b^3$, 4) $24x^8$, 5) x^{12}, 6) a^{23}

Name:

Instructor:

Date:

Section:

Examples 10.2 (cont'd)

Objective C Use the power of a product rule for exponents.

Example 7 Simplify: $\left(5x^3\right)^3$

Example 8 Simplify: $\left(3a^2b^3\right)^4$

Example 9 Simplify: $\left(2x^2y^3\right)^2\left(3x^2y^4z\right)^4$

7) $125x^9$, 8) $81a^8b^{12}$, 9) $324x^{12}y^{22}z^4$

Martin-Gay **Prealgebra** edition 5

362

Practice Set 10.2

Objective A Use the product rule for exponents.

Multiply.

1. $x^5 \cdot x^6$

1. _____

2. $2x^3 \cdot 3x^5$

2. _____

3. $2x \cdot 3x \cdot 4x$

3. _____

4. $4y^2 \cdot 2y^2 \cdot 2y$

4. _____

5. $\left(-2x^3 y^6\right)\left(-3x^4 y^3\right)$

5. _____

6. $\left(-2x^3 y^6\right)\left(-2x^3 y^4\right)$

6. _____

Objective B Use the power rule for exponents.

7. $\left(x^3\right)^6$

7. _____

8. $\left(y^4\right)^9$

8. _____

9. $\left(a^5\right)^8$

9. _____

10. $\left(z\right)^{20}$

10. _____

11. $\left(x^3\right)^8 \cdot \left(x^2\right)^5$

11. _____

12. $\left(3a^4\right)^4$

12. _____

Practice Set 10.2 (cont'd)

Objective C Use the power of a product rule for exponents.

13. $\left(x^2 y^4\right)^6$ 13. _____

14. $\left(a^3 b^2\right)^5$ 14. _____

15. $\left(x^3 y^5\right)^6$ 15. _____

16. $\left(5x^3 b^8\right)^2$ 16. _____

17. $(-3y)^4 \left(3y^2\right)^2$ 17. _____

18. $(3xy)^4 \left(2x^3 y^5 z^3\right)^3$ 18. _____

19. $(2xy)^4 \left(3x^5 y^6\right)^2$ 19. _____

Extensions

20. The area of the triangle.   20.

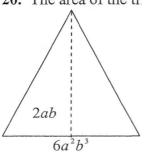

10.3 Multiplying Polynomials

Learning Objectives
A. Multiply a monomial and any polynomial.
B. Multiply two binomials.
C. Square a binomial.
D. Use the FOIL order to multiply binomials.
E. Multiply any two polynomials.

Objective A Multiply a monomial and any polynomial.

Example 1 Multiply $3x(2x^2 - 4)$

Example 2 $3y(2y^2 - 4y + 6)$

Objective B Multiply two binomials.

Example 3 Multiply: $(x + 4)(x + 5)$

Example 4 Multiply: $(2y - 3)(3y + 4)$

Objective C Square a binomial.

Example 5 $(3x - 2)^2$

1) $6x^3 - 12x$, 2) $6y^3 - 12y^2 + 18y$, 3) $x^2 + 9x + 20$, 4) $6y^2 - y - 12$, 5) $9x^2 - 12x + 4$

Examples 10.3 (cont'd)

Objective D Use the FOIL order to multiply binomials.

Use the FOIL order to multiply.

   Example 6 $(3x-2)(4x+3)$

   Example 7 $(5x-6)^2$

Objective E Multiply any two polynomials.

   Example 8 Multiply: $(3x-4)(x^2-4x+3)$

   Example 9 Find the product of (a^2-3a-4) and $(2a-3)$ vertically.

6) $12x^2+x-6$, 7) $25x^2-60x+36$, 8) $3x^3-16x^2+25x-12$, 9) $2a^3-9a^2+a+12$

Practice Set 10.3

Objective A Multiply a monomial and any polynomial..

Muliplty.

1. $2x(7x^2 - 5)$

1. _____

2. $-4ab(2b^2 - 4a + 2)$

2. _____

3. $-3a^2(2a^2 - 4ab + 3b^2)$

3. _____

Objective B Multiply two Binomials.

Multiply.

4. $(y - 3)(y + 2)$

4. _____

5. $(2x - 3)(4x + 2)$

5. _____

6. $(3x - 8)(2x - 5)$

6. _____

7. $(5x - 1)(5x + 1)$

7. _____

Objective C Square a Binomial

8. $(4x - 5)^2$

8. _____

9. $(7b - 3)^2$

9. _____

10. $(2x + 5)^2$

10. _____

Objective D Use the FOIL order to multiply binomials.

Multiply.

11. $(3x - 7)(2x + 4)$

11. _____

12. $(4x - 5)(3x + 2)$

12. _____

13. $(3x - 5)(7x - 2)$

13. _____

Practice Set 10.3 (cont'd)

14. $(2x-5)(2x+5)$ 14. _____

Objective E Multiply any two polynomials.

Multiply.

15. $(x+2)(x^2-3x+5)$ 15. _____

16. $(2x-3)(3x^2-5x+7)$ 16. _____

17. $(y-3)(y^2-2y+3)$ 17. _____

18. $(x-3)(x^2+3x+9)$ 18. _____

19. $(2x^2-2x+3)(x^2-5x-7)$ 19. _____

Extensions

20. Find the area of the shaded figure.   20. _____

$(x - 1)$ inches

$(2x - 11)$ inches

10.4 Introduction to Factoring Polynomials

Learning Objectives
A. Find the greatest common factor of a list of integers.
B. Find the greatest common factor of a list of terms.
C. Find the greatest common factor from the terms of a polynomial.

Objective A Find the greatest common factor of a list of integers.

Example 1 Find the GCF of 16 and 24.

Objective B Find the greatest common factor of a list of terms.

Example 2 Find the GCF of x^4, x^7, and x^{12}.

Example 3 Find the GCF of $3x^2$, $9x$, and $12x^4$.

1) 8, 2) x^4 3) 3x,

Examples 10.4 (cont'd)

Objective C Factor the greatest common factor from the terms of a polynomial.

    Example 4 Factor. $6x^3 + 12x^2$

    Example 5 $3x^2 - 12x + 18$

    Example 6 Factor two ways. $-5a + 20b + 10b^2$

4) $6x^2(x + 2)$, 5) $3(x^2 - 4x + 6)$ 6) $5(-a + 4b + 2b^2)$ or $-5(a - 4b - 2b^2)$

Practice Set 10.4

Objective A Find the greatest common factor of a list of integers.

Find the greatest common factor of each list of integers.

1. 42 and 36

1. _____

2. 50 and 15

2. _____

3. 112 and 54

3. _____

4. 12, 18, and 24

4. _____

5. 27, 81 and 90

5. _____

6. 30, 50, and 100

6. _____

Objective B Find the greatest common factor of a list of terms.

Find the greatest common factor of each list of terms.

7. y^5, y^7, and y^{12}

7. _____

8. b^5, b^8, and b

8. _____

9. x^2y, x^5y^3, and x^7y^{19}

9. _____

10. $2x^3$, $4x^5$, and $6x^7$

10. _____

11. $8x^3$, $12x^5$, and $24x^9$

11. _____

Practice Set 10.4 (cont'd)

12. $6x^5$, $15x^7$, and $21x^3$ 12. _____

Objective C Find the greatest common factor from the terms of a polynomial.

Factor.

13. $4x^2 - 8x$ 13. _____

14. $9y^2 - 7y$ 14. _____

15. $a^5 - 3a^4$ 15. _____

16. $9x^2 - 12$ 16. _____

17. $4x^3 - 2x^2 - 8x$ 17. _____

18. $4x^3 - 12x^5$ 18. _____

19. Factor two ways. $-3a^3 - 6a^2 + 12a$ 19. _____

Extensions

20. The missing side of the rectangle if the area of   20. _____
the rectangle is $6x^2 + 12x$.

6x inches
?

Chapter 10 Vocabulary Reference Sheet

Term	Definition	Example
	Section 10.1	
Monomial	A term that contains whole number exponents and no variable in the denominator.	$15x^3 y^5$ is a monomial.
Polynomial	A monomial or a sum or difference of monomials.	$9x^2 - 5x + 18$, $8x$, and $\dfrac{2}{3}x$ are polynomials.
Binomial	A polynomial with two terms.	$3x - y$ is a binomial.
Trinomial	A polynomial with three terms.	$3x^2 - 5x + 12$ is a trinomial.

NOTES:

Chapter 10 Practice Test A

Add or subtract as indicated.

1. $(3x - 2) + (4x - 6)$ 1. _____

2. $(2.3x^2 + 5x) + (1.7x^2 - 3x)$ 2. _____

3. $(10x - 7) - (8x - 9)$ 3. _____

4. Subtract $(9x^2 - x)$ from $(6x^2 - 4x - 9)$.   4. _____

5. Find the value of $x^2 - 3x + 6$ when $x = -2$.   5. _____

Multiply and simplify.

6. $x^4 \cdot x^9$ 6. _____

7. $(a^5)^7$ 7. _____

8. $(3x^2)^3$ 8. _____

9. $(3x^2)(-2x^3)$ 9. _____

10. $(b^3)^5 (b^5)^8$ 10. _____

11. $(3ab^2c^3)^2 (2a^3bc^5)^4$ 11. _____

12. $3x(5x - 9)$ 12. _____

13. $-5x(x^2 - 7x + 9)$ 13. _____

14. $(x - 5)(x + 4)$ 14. _____

15. $(3x - 2)(3x - 5)$ 15. _____

Chapter 10 Practice Test A (cont'd)

16. $(3x - 5)(3x + 5)$

16. _____

17. $(4x - 9)^2$

17. _____

18. $(a - 3)(a^2 - 4a - 9)$

18. _____

19. Find the area and the perimeter of the parallelogram.
(*Hint:* $A = b \cdot h$.)

19. _____

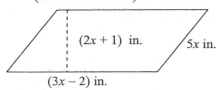

(2x + 1) in. 5x in.

(3x − 2) in.

Find the greatest common factor of each list.

20. 18 and 36

20. _____

21. $4y^2$, $8y^5$, and $16y^9$

21. _____

Factor out the GCF.

22. $5y^2 - 15y$

22. _____

23. $14a^2 - 7a$

23. _____

24. $9x^2 - 6x - 12$

24. _____

25. $12x^7 - 11x^6 + 7x^5$

25. _____

Chapter 10 Practice Test B

Add or subtract as indicated.

1. $(5x - 7) + (3x - 2)$
   a. $8x - 5$ **b.** $8x - 9$ **c.** $8x - 3$ **d.** $6x - 2$

2. $(1.9x^2 - 3x) + (0.6x^2 + 5x)$
   a. $7.9x^2 + 2x$ **b.** $2.5x^2 - 2x$ **c.** $2.5x^2 + 2x$ **d.** $1.3x^2 + 8x$

3. $(3x - 5) - (5x - 7)$
   a. $-2x + 2$ **b.** $-2x - 12$ **c.** $8x - 12$ **d.** $8x + 2$

4. Subtract $(4x^2 - 2x)$ from $(5x^2 + 2x - 3)$.
   a. $-x^2 - 4x + 3$ **b.** $-x^2 - 3$ **c.** $x^2 + 4x - 3$ **d.** $x^2 - 3$

5. Find the value of $x^2 - 5x - 6$ when $x = -3$.
   a. -12 **b.** -30 **c.** 0 **d.** 18

Multiply and simplify.

6. $x^3 \cdot x^5$
   a. x^8 **b.** x^{15} **c.** x^2 **d.** x^{12}

7. $(a^3)^5$
   a. a^8 **b.** a^{15} **c.** a^9 **d.** a^{19}

Chapter 10 Practice Test B

8. $\left(-2x^3\right)^2$

    a. $-2x^5$ **b.** $-2x^6$ **c.** $4x^6$ **d.** $4x^5$

9. $\left(4x^2\right)\left(3x^5\right)$

a. $12x^{10}$ **b.** $7x^7$ **c.** $7x^{10}$ **d.** $12x^7$

10. $\left(b^2\right)^4\left(b^4\right)^6$

    a. b^{192} **b.** b^{32} **c.** b^{16} **d.** a^{22}

11. $\left(2a^2b^3\right)^3\left(-4a^2bc^3\right)^2$

    a. $-8a^{10}b^9c^6$ **b.** $-8a^{10}b^{11}c^6$ **c.** $128a^{10}b^{11}c^6$ **d.** $128a^{10}b^9c^{15}$

12. $4x(3x-7)$

    a. $12x^2-7$ **b.** $12x-28x$ **c.** $12x^2-28$ **d.** $12x^2-28x$

13. $-2x\left(2x^2-7x-3\right)$

    a. $-4x^3+14x^2+6x$ **b.** $-4x^3-7x-3$

    c. $-4x^2-14x-6x$ **d.** $-4x^2+17x+6x$

14. $(x-7)(x+8)$

    a. x^2-56 **b.** x^2+x-56 **c.** x^2-x-56 **d.** $x^2-15x+56$

15. $(4x-2)(3x-7)$

    a. $12x^2+14$ **b.** $12x^2-22x+14$ **c.** $12x^2+22x-14$ **d.** $12x^2-34x+14$

Chapter 10 Practice Test B

16. $(4x - 7)(4x + 7)$

    a. $16x^2 + 49$ **b.** $16x^2 - 56x + 49$ **c.** $16x^2 - 49$ **d.** $16x^2 - 56x - 49$

17. $(3x - 9)^2$

    a. $3x^2 - 81$ **b.** $9x^2 - 81$ **c.** $9x^2 - 54x - 81$ **d.** $9x^2 - 54x + 81$

18. $(a + 2)(2a^2 - 3a + 5)$

    a. $2a^3 + a^2 - a + 10$ **b.** $2a^3 - 3a^2 - 6a + 10$

    c. $2a^3 + a^2 + a + 10$ **d.** $2a^3 - a^2 + 11a + 10$

19. Find the area of the parallelogram.
    (*Hint:* $A = b \cdot h$.)

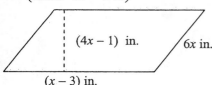

    a. $6x^2 - 18x$ **b.** $4x^2 - 13x + 3$ **c.** $24x^2 - 6$ **d.** $24x^2 - 1$

Find the greatest common factor of each list.

20. 30 and 45

    a. 5   **b.** 3   **c.** 15   **d.** 30

21. $5y^2$, $10y^5$, and $25y^8$

    a. $5y$ **b.** $5y^2$ **c.** $25y^8$ **d.** $10y^5$

Chapter 10 Practice Test B

Factor out the GCF.

22. $4y^2 - 6y$
    a. $2y(2y - 3)$ **b.** $y(4y - 6)$ **c.** $2(2y^2 - 3y)$ **d.** $2y(2y - 6y)$

23. $25a^2 - 5a$
    a. $5(5a^2 - a)$ **b.** $a(25a - 5)$ **c.** $5a^2(5a)$ **d.** $5a(5a - 1)$

24. $8x^2 - 16x + 24$
    a. $4(2x^2 - 8x + 6)$ **b.** $4(2x^2 - 8x + 6)$
    c. $8(x^2 - 2x + 3)$ **d.** $x(8x - 16x + 24)$

25. $21x^9 - 14x^7 + 7x^3$
    a. $7x^3(3x^6 - 2x^4 + 1)$ **b.** $7x^3(3x^2 - 2x^4)$
    c. $7(3x^9 - 2x^7 + x^3)$ **d.** $7x(3x^8 - 2x^6 + x^2)$

Chapter 1

Practice Set 1.2
1. words
3. expanded form
5. place value
7. ten
9. ten thousand
11. twenty-three thousand, eight hundred thirty-one
13. thirteen thousand, one hundred twenty-six
15. 9816
17. $10,000 + 3000 + 200 + 9$
19. $10,000,000 + 2,000,000 + 300,000 + 9000 + 100 + 40 + 1$
21. 135
23. 9999
25. no, two hundred six

Practice Set 1.3
1. addend, sum
3. minuend, subtrahend, difference
5. associative
7. 6203
9. 29
11. 12,899
13. 34 yards
15. 96
17. $389
19. $515
21. 89, 184 minuend, 95 subtrahend, 89 difference

Practice Set 1.4
1. exact, estimate
3. 400
5. 70
7. 4390; 4400; 4000

Practice Set 1.4 (cont)
9. 28,950; 29,000; 29,000
11. $250,000
13. 190
15. 0
17. $1180
19a. 2650
   b. 2749

Practice Set 1.5
1. product, factor
3. associative
5. distributive
7. 23
9. 0
11. $12 \cdot 9 + 12 \cdot 4$
13. 3856
15. 1,157,958
17. 875 square feet
19. 1242
21. $3024
23. 6400

Practice Set 1.6
1. undefined
3. 6
5. 9
7. 42
9. 0
11. 28
13. 75
15. 18 R 2
17. 62 R 8
19. 583 R 8
21. $185,000
23. 24
25. answers will vary

Answers

Practice Set 1.7
1. base, exponent
3. 12^4
5. 8
7. 12
9. 36
11. 14
13. 11
15. 12
17. 104
19. 3
21. 10
23. 100 square feet
25. 70 inches

Practice Set 1.8
1. variable
3. variable
5. 14
7. 38
9. 5
11. 41
13. 81
15. $4 + x$
17. $x - 7$
19. $x - 5$

Chapter 1

Answers

(cont'd)

Test A

1. eight hundred seventy-six thousand, one hundred twenty-four
2. 72,008
3. 226
4. 105
5. 9300
6. 133 R 26
7. 432
8. 15
9. 0
10. undefined
11. 1
12. 22
13. 80
14. 35
15. 1
16. 59
17. 160,000
18. 11,800
19. 4
20. 59
21. 84
22. 17
23. $18
24. $35
25. $196
26. $907
27. area = 64 square feet, perimeter = 32 feet
28. area = 96 square feet, perimeter = 40 feet
29. 12
30. 2
31. $10 + x$
32. $6x$
33. yes

Test B

1. b
2. a
3. d
4. b
5. d
6. a
7. d
8. b
9. a
10. c
11. c
12. a
13. b
14. d
15. c
16. b
17. a
18. b
19. b
20. c
21. a
22. b
23. a
24. c
25. d
26. b
27. c
28. a
29. d
30. c
31. b
32. a
33. b

Chapter 2

ANSWERS

Practice Set 2.1
1. integers
3. signed numbers
5. −8
7.
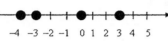
9. >
11. 4
13. 0
15. 6
17. 18
19. −6
21. >
23. −4, −(2), 1, −(−3), |−4|
25. answers will vary

Practice Set 2.2
1. 0
3. −3
5. 4
7. −3
9. −2
11. −4
13. 0
15. 3
17. −3500
19. answers will vary

Practice Set 2.3
1. −3
3. 4
5. 6
7. −3
9. −3
11. −9
13. −2
15. 3
17. −2
19. 5
21. 8 points
23. answers will vary

Practice Set 2.4
1. negative
3. negative
5. 0
7. 18
9. 0
11. 60
13. −16
15. 4
17. −6
19. −8
21. 48
23. −12 yards
25. positive
27. −1

Practice Set 2.5
1. multiplication
3. addition
5. 1
7. 4
9. 12
11. −1
13. −56
15. −36
17. −2
19. 8

Practice Set 2.6
1. expression, equation
3. multiplication
5. expression, equation
7. no
9. −14
11. −5
13. 7
15. 8
17. −8
19. 200
21. answers will vary

Chapter 2

Answers

Test A

1. −1
2. −7
3. −10
4. 4
5. −6
6. −1
7. −60
8. −5
9. 29
10. 11
11. 6
12. −1
13. −8
14. −7
15. −12
16. 256
17. 2
18. −10
19. −2
20. −5
21. −12
22. 0
23. 1
24. $445
25. −67 feet
26. 43,973 feet
17. 2
28. −5
29. 32
30. −2
31. −2

Test B

1. b
2. c
3. a
4. b
5. d
6. c
7. a
8. b
9. a
10. a
11. d
12. b
13. d
14. a
15. d
16. b
17. c
18. b
19. c
20. a
21. d
22. a
23. c
24. d
25. b
26. a
27. c
28. a
29. d
30. c
31. b
32. a
33. d

Chapter 3

Answers

Practice Set 3.1

1. combine like terms
3. coefficient
5. $10x$
7. $-y + 3$
9. $-6x$
11. $3x + 12$
13. $2x - 2$
15. $8x - 13$
17. $-12b + 8$
19. $19x + 26$ feet
21. 700 square feet
23. $86x$ square feet

Practice Set 3.2

1. simplify
3. 10
5. 5
7. -3
9. 4
11. -20
13. 0
15. $x + 9$
17. $-3x$
19. $8x + 25$

Practice Set 3.3

1. 10
3. 0
5. -3
7. 3
9. -8
11. -8
13. -10
15. 0
17. $-8 - 12 = -20$
19. $4(8 - 5) = 12$
21. 2

Practice Set 3.4

1. $x - 4 = 8$
3. $8 - x = -25$
5. $2x - 8 = 24$
7. 3
9. 24
11. 7
13. 1a
15. Ruth 60 mph, Frank 65 mph
17. 174 cars
19. $18

Test A

1. $-4x + 16$
2. $-8x + 12$
3. $2x - 6$
4. $8x + 8$
5. $20x - 15$
6. -1
7. 9
8. -1
9. 7
10. -2
11. -9
12. 12
13. 7
14. -3
15. $-6 + x$
16. $7 - 4x$
17. $3x + 9 = -7$
18. $4x + 6 = 25$
19. 8
20. 16 males, 21 females

Test B

1. b
2. d
3. a
4. c
5. a
6. b
7. a
8. d
9. a
10. c
11. b
12. c
13. a
14. d
15. b
16. a
17. c
18. d
19. b
20. b

Chapter 4

Practice Set 4.1

1. fraction, numerator, denominator
3. mixed number
5. $\dfrac{5}{7}$
7.
9. $\dfrac{271}{519}$
11.
13. 1
15. 0
17. $\dfrac{26}{7}$
19. $\dfrac{38}{3}$
21. $5\dfrac{3}{10}$
23. 1
25.

Practice Set 4.2

1. $2^3 \cdot 3$
3. $2 \cdot 5 \cdot 7$
5. $\dfrac{2}{9}$
7. $-\dfrac{2}{7}$
9. $\dfrac{7}{9xy}$
11. no
13. no
15. $\dfrac{1}{3}$
17. $\dfrac{28}{43}$
19. $\dfrac{3}{4}$

Answers

Practice Set 4.3

1. multiplication
3. $\dfrac{6}{35}$
5. $-\dfrac{3}{10}$
7. $\dfrac{1}{2x}$
9. $-\dfrac{1}{8}$
11. $\dfrac{4x}{5}$
13. $\dfrac{5}{9}$
15. $\dfrac{1}{16}$
17. $\dfrac{1}{3}$
19. 40
21. 300 calories
23. $\dfrac{25}{36}$, 1, yes

Practice Set 4.4

1. unlike, like
3. equivalent
5. $1\dfrac{1}{6}$
7. $\dfrac{1}{15}$
9. $\dfrac{1}{5}$
11. $\dfrac{4x+1}{3}$
13. 1
15. $\dfrac{6}{7}$ inch
17. 15
19. 30
21. $\dfrac{16}{28}$

Practice Set 4.4 con't

23. $\dfrac{1}{8}$

Practice Set 4.5

1. $\dfrac{17}{20}$
3. $\dfrac{19y}{21}$
5. $\dfrac{8b-1}{6}$
7. $-\dfrac{13}{20}$
9. $-\dfrac{17}{21}$
11. $-\dfrac{1}{5}$
13. <
15. $-\dfrac{11}{12}$
17. $-1\dfrac{1}{6}$
19. $\dfrac{1}{2}$ in

Practice Set 4.6

1. $\dfrac{\frac{1}{2}}{3}$
3. $1\dfrac{1}{3}$
5. $\dfrac{6x}{7}$
7. 2
9. $\dfrac{2}{3}$
11. $-\dfrac{56}{225}$
13. $\dfrac{1}{36}$
15. $\dfrac{1}{9}$
17. $\dfrac{25}{36}$
19. $\dfrac{5}{9}$

Chapter 4

Practice Set 4.7

1. mixed number
3. improper
5. $\dfrac{7}{10}$
7. 26
9. $\dfrac{11}{16}$
11. $14\dfrac{2}{21}$
13. $2\dfrac{3}{4}$
15. $24\dfrac{3}{8}$
17. $9\dfrac{3}{4}$
19. $-6\dfrac{27}{35}$

Practice Set 4.8

1. $\dfrac{19}{20}$
3. $\dfrac{19}{35}$
5. $-\dfrac{1}{2}$
7. $\dfrac{1}{2}$
9. $10\dfrac{1}{2}$
11. $\dfrac{2}{3}$
13. 9
15. $\dfrac{3x-4}{6}$
17. $\dfrac{9+5x}{3}$
19. $3\dfrac{1}{2}$ inch

Answers

Test A

1. $\dfrac{4}{5}$
2. $\dfrac{19}{3}$
3. $3\dfrac{1}{9}$
4. $\dfrac{4}{5}$
5. $\dfrac{5x}{8}$
6. yes
7. $2^5 \cdot 3$
8. $\dfrac{1}{2}$
9. $-\dfrac{5}{9}$
10. $\dfrac{19x}{28}$
11. $\dfrac{2x-15}{5x}$
12. $\dfrac{5}{14}$
13. $-\dfrac{1}{3}$
14. $\dfrac{1}{4}$
15. $-\dfrac{26}{35}$
16. $\dfrac{2}{7}$
17. 0
18. $2\dfrac{1}{5}$
19. $2\dfrac{2}{3}$
20. $1\dfrac{10}{21}$
21. $-\dfrac{1}{30}$
22. $-\dfrac{7}{36}$
23. $\dfrac{2}{3}$
24. 3
25. $-1\dfrac{2}{7}$
26. $1\dfrac{14}{15}$
27. $-\dfrac{1}{4}$
28. 0
29. $1\dfrac{1}{3}$
30. $2\dfrac{1}{2}$ feet
31. $2\dfrac{1}{6}$ feet
32. $\dfrac{1}{4}$ square foot
33. 8 pieces

Chapter 4 Test B

1. b
2. c
3. d
4. a
5. c
6. b
7. c
8. d
9. a
10. b
11. c
12. a
13. d
14. a
15. d
16. c
17. b
18. a
19. c
20. b
21. d
22. c
23. a
24. a
25. b
26. b
27. a
28. b
29. c
30. a
31. d
32. a
33. c

Chapter 5

Practice Set 5.1

1. six and forty-eight hundredths
3. negative seventeen hundredths
5. one hundred six and four tenths
7. 5.08
9. −1.06
11. $\frac{9}{10}$
13. $\frac{4}{25}$
15. $3\frac{1}{250}$
17. >
19. 0.37
21. −2.05
23. $38
25. 0.001, 0.047, 0.072, 0.135, 0.136

Practice Set 5.2

1. 4.83
3. 1.81
5. −8.93
7. 9.3
9. 62.81
11. −2.33
13. 5.1
15. no
17. $9.8x − 5.8$
19. 6.52 inches
21. 1.37 inches

Practice Set 5.3

1. sum
3. left
5. 0.15
7. 0.244
9. 0.177
11. 0.6
13. 0.00467
15. −3.68
17. 24π inches, 75.36 inches
19. 4.8 grams

Practice Set 5.4

1. left
3. 2.39
5. 3.2
7. 50,000
9. 70
11. 2.6
13. 0.03621
15. −0.0135
17. 2
19. 12.7

5.5

1. 0.48
3. $0.41\overline{6}$
5. 3.6
7. 4.43
9. >
11. 0.411, 0.43, $\frac{11}{25}$
13. −12
15. 7.25 square inches
17. −8.01
19. −10.2
21. 0.15

Practice Set 5.6

1. −8.1
3. 3.5
5. −8.8
7. −1.27
9. 2.4
11. 0.2
13. 7.78
15. −1.35
17. −46.5
19. −1.51

5.7

1. mode
3. median
5. 24
7. 515.125
9. 2.85
11. 27
13. 500.5
15. 27
17. no mode
19. 118, 121

Chapter 5

Test A

1. thirty-six and one hundred ninety five thousandths
2. 1.065
3. 8.215
4. −4.36
5. −13.733
6. 8.673
7. −23.1
8. 47.3
9. 0.92
10. <
11. >
12. $\dfrac{19}{25}$
13. $-5\dfrac{6}{25}$
14. −0.52
15. 0.69
16. −3.21
17. −13
18. $2.9x + 7.78$
19. 2.4
20. 1.08
21. 40.8, 47, no mode
22. 35, 42, 48
23. 2.93
24. 9.3 square inches
25. 43.96 yards

Answers

Test B

1. b
2. c
3. c
4. a
5. b
6. c
7. d
8. c
9. d
10. b
11. b
12. c
13. d
14. a
15. d
16. c
17. d
18. d
19. a
20. b
21. c
22. b
23. a
24. b
25. c

Chapter 6

Practice Set 6.1

1. unit
3. ratio
5. $\dfrac{1}{3}$
7. $\dfrac{11}{4}$
9. $\dfrac{6}{11}$
11. $\dfrac{1 \text{ table}}{4 \text{ diners}}$
13. 4 books/student
15. 2.4 pages/minute
17. $0.20 per banana
19. $2.40 for 12 pens

Practice Set 6.2

1. proportion
3. cross product
5. $\dfrac{3\frac{1}{2}}{7} = \dfrac{7\frac{1}{2}}{15}$
7. false
9. false
11. false
13. −20
15. −0.9
17. 4.7
19. answers will vary

Practice Set 6.3

1. 10 attempts
3. 40 accepted
5. 160 miles
7. 585.6 miles
9. 16,800 square feet
11. 28 hits
13. 1 cup
15. 18 cups of flour
17. 260 pens
19. 870 mg

Chapter 6

Practice Set 6.4

1. square root
3. perfect squares
5. $x^2 + y^2 = z^2$
7. 5
9. $\frac{6}{7}$
11. 3.873
13. 4.899
15. 11.136
17. 13
19. 9 and 10

Practice Set 6.5

1. similar
3. yes
5. yes
7. $\frac{4}{3}$
9. $\frac{3}{2}$
11. 11.25
13. 8
15. 8.3 inches
17. 64 feet
19. answers will vary

Answers

Test A

1. $\frac{12}{17}$
2. 65 miles/hour
3. $\frac{2}{5}$
4. 2
5. $\frac{7}{9}$
6. 45 miles/gallon
7. 5 miles/hour
8. 15 applicants/job
9. 16 ounces for $2.06
10. 12 ounces for $1.21
11. true
12. false
13. 6
14. $4\frac{1}{2}$
15. 4
16. 3
17. 5 inches
18. 5 hours
19. 750 ml
20. 9
21. 12.767
22. $\frac{7}{10}$
23. 7.81
24. 9 inches
25. 33 feet

Test B

1. b
2. a
3. c
4. b
5. d
6. b
7. c
8. d
9. a
10. a
11. b
12. a
13. d
14. c
15. d
16. a
17. b
18. a
19. c
20. d
21. b
22. b
23. c
24. a
25. c

Chapter 7

Practice Set 7.1
1. per hundred
3. 13%
5. 0.21
7. 0.034
9. $\dfrac{3}{50}$
11. $\dfrac{7}{200}$
13. 50%
15. 0.6%
17. 60%
19. 375%
21. 42.86%
23. 4%
25. <

Practice Set 7.2
1. is
3. $0.15x = 80$
5. $120x = 4$
7. $x = 0.1 \cdot 180$
9. $180x = 200$
11. 12
13. 140
15. 76%
17. 24
19. 50%

Practice Set 7.3
1. percent, base, amount
3. base
5. $\dfrac{x}{80} = \dfrac{72}{100}$
7. $\dfrac{12.9}{x} = \dfrac{15}{100}$
9. $\dfrac{70}{490} = \dfrac{x}{100}$
11. 210
13. 200%
15. 150
17. 75
19. 28

Practice Set 7.4
1. 640 bulbs
3. 18%
5. 400 televisions

Practice Set 7.4 con't
7. $600
9. 4, 20%
11. 3, 30%
13. 27, 18%
15. 20%
17. 10%
19. 15%

Practice Set 7.5
1. $55.25
3. $150, $160.50
5. $450, $477
7. $19,000
9. $45,000
11. $12,000
13. $7.36, $84.64
15. $45, $105
17. $300
19. $54.06

Practice Set 7.6
1. simple
3. compound
5. $216, $1016
7. $312, $1512
9. $70, $2870
11. $180, $1380
13. $16,325.87
15. $4666.36
17. $1779.04
19. $12,600.66

Test A
1. 0.32
2. 1.5
3. 0.0005
4. 60%
5. 320%
6. 1.2%
7. $\dfrac{9}{25}$
8. $1\dfrac{2}{5}$
9. $\dfrac{1}{2500}$
10. 28%
11. 85%
12. 162.5%

Test A con't
13. 25%
14. $\dfrac{2}{25}$
15. 168
16. 2000
17. 25%
18. 9 ml
19. $250,000
20. $617.70
21. 12%
22. $270
23. $720
24. 8%
25. $675
26. $2380.31
27. $1534.95

Test B
1. b
2. c
3. d
4. c
5. a
6. b
7. a
8. c
9. d
10. b
11. a
12. c
13. a
14. b
15. d
16. a
17. c
18. b
19. a
20. d
21. c
22. a
23. b
24. c
25. d
26. a
27. b

Chapter 8

8.1

1. pictograph
3. bars
5. Texas
7. $600,000,000
9. 475 calories
11.

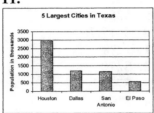

13. 15 students
15. Class Frequency

    4

    8

    4

    2

    2

17. $100,000
19. $200,000,000

8.2

1. 360°
3. shoes
5. $\frac{2}{3}$
7. $\frac{3}{4}$
9. 40%
11. 300 books
13. 30%
15. action
17. $\frac{2}{3}$
19.

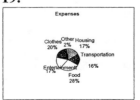

Answers

8.3

1. x, y
3. quadrants
5. plane

7.

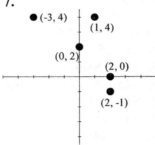

9. F (−2, 3), G (0, −3),
   H 1, 2), I (2, 0),
   J (2, −3
11. no
13. no
15.

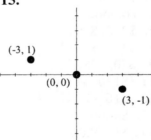

17. (3, 2), (2, 0),
   (4, 4)
19. (4, 1), (5, 2),
   (0, −3)

8.4

1. horizontal
3.

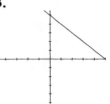

5.

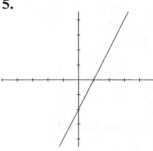

7.

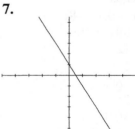

9.

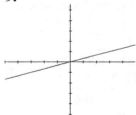

11.

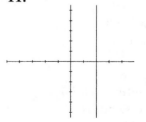

Chapter 8

8.4 (cont'd)

13.

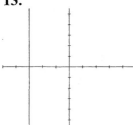

15.

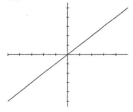

17.

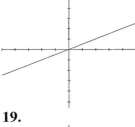

19.

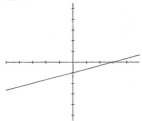

8.5

1. probability

3. 1

5.

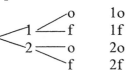

	o	1o
1	f	1f
2	o	2o
	f	2f

7.

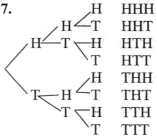

	H	HHH
H—T		HHT
H—T—H		HTH
	T	HTT
	H	THH
T—H—T		THT
T—H		TTH
	T	TTT

9. $\dfrac{1}{3}$

11. 0

13. $\dfrac{3}{13}$

15. $\dfrac{9}{13}$

17. $\dfrac{1}{10}$

19. $\dfrac{1}{8}$

Test A

1. January

2. 200 puppies

3. 50 puppies

4. 2 weeks

5. 8000

6. 4th week

7.

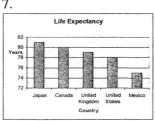

Life Expectancy

8. 10,000

9. 2005 to 2006

10. 8000

11. $\dfrac{3}{4}$

12. $\dfrac{2}{3}$

13. 160

14. 120

15. 12

16. 8

Test A con't

17. Class Frequencies

3
2
6
5
5
1

18.

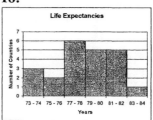

Life Expectancies

19. (3, 1)

20. (0, −2)

21. (−1, 2)

22. (−2, −3)

23.

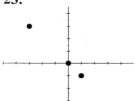

24.

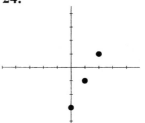

25.

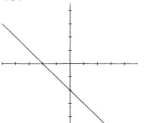

26.

27.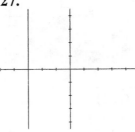

28.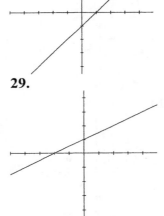

29.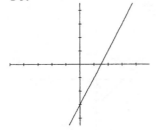

30.

Test A con't

31.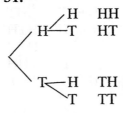

H   HH
H—T   HT

T—H   TH
   T   TT

32. $\frac{1}{3}$

33. $\frac{1}{4}$

Test B

1. a
2. c
3. a
4. c
5. a
6. c
7. c
8. c
9. a
10. d
11. b
12. c
13. a
14. c
15. a
16. a
17. c
18. c
19. b
20. c
21. d
22. c
23. a
24. c
25. b
26. d
27. c
28. a
29. d
30. b
31. c
32. b
33. a

Chapter 9

9.1
1. adjacent
3. ray
5. obtuse
7. line, $\overrightarrow{TS}$
9. angle, $\vee$ PQR
11. $\vee$BAC, $\vee$CAB
13. straight
15. obtuse
17. 123°
19. 119°

9.2
1. circumference
3. diameter
5. 44 yd
7. 22 yd
9. 31 yd
11. 60 yd
13. 26 in.
15. 60 cm
17. $18\pi \approx 56.52$ in.
19. $7\pi \approx 21.98$ in.

9.3
1. square
3. square
5. volume
7. 28 sq in.
9. $201\dfrac{1}{7}$ sq ft
11. 114 sq ft
13. 480 sq ft
15. 729 cu cm, 486 sq cm
17. $38\dfrac{2}{21}$ cu in., $523\dfrac{17}{21}$ sq in.
19. 75 cu ft

9.4
1. 6 ft
3. 5 in.
5. 6 yd 2 ft
7. 17 ft
9. 8 ft 2 in.
11. 6 ft 10 in.
13. 4800 cm
15. 56.4 m
17. 8mm
19. 2.1 km

9.5
1. mass
3. 16
5. 48 oz
7. $2\dfrac{1}{4}$ tons
9. 56 oz
11. 2 lb 12 oz
13. 1 ton 800 lb
15. 50,000 g
17. 7210 mg
19. 1.28 g or 1280 mg

9.6
1. capacity
3. 2 c
5. 3 gal
7. 12 gal 2 qt
9. 6 c 3 oz
11. 10 gal 2 qt
13. 9 ml
15. 140 L
17. 48.9 L
19. 3000 ml

9.7
1. 27.05 fl oz
3. 228.6 cm
5. 20 oz
7. 66.96 mi
9. 3.25 gal
11. 27 kg
13. 41°F
15. 190.4°F
17. 4.4°C
19. 13.3°C

Test A
1. 72°
2. 159°
3. 30°
4. 48°
5. 125°
6. 12 in.
7. 12 ft
8. C = 75.36 in.
   A = 452.16 sq in.
9. P = 4.96 ft
   A = 1.47 sq ft
10. P = 100 ft
    A = 568 sq ft
11. $75\dfrac{3}{7}$ cu in.
12. 192 cu in.
13. 48 in.
14. 210 cu ft
15. 64 ft, $224
16. 5 ft 8 in
17. 6 qt
18. 5 lb
19. 6400 lb
20. $3\dfrac{3}{4}$ gal
21. 0.03 g
22. 1400 g
23. 21 mm
24. 0.8 g
25. 930 ml
26. 8 qt
27. 2 lb 10 oz
28. 11 ft
29. 1 gal 1 qt
30. 2.098 km or 2098 m
31. 0.4 cm or 4 mm
32. 23.9°C
33. 65.5°F

Chapter 9

1. b
2. c
3. d
4. a
5. a
6. c
7. d
8. b
9. d
10. b
11. c
12. d
13. a
14. d
15. c
16. b
17. a
18. b
19. c
20. d
21. a
22. b
23. d
24. c
25. c
26. a
27. d
28. b
29. a
30. b
31. c
32. d
33. a

Chapter 10

10.1
1. trinomial
3. monomial
5. $-2x - 9$
7. $7x^2 + 2x - 1$
9. $x^2 - 12x - 1$
11. $-3x^2 + 7x + 9$
13. $-x^2 - 3x - 12$
15. $-2a^2 - 5a + 18$
17. 23
19. 4

10.2
1. x^{11}
3. $24x^3$
5. $6x^7 y^9$
7. x^{18}
9. a^{40}
11. x^{34}
13. $x^{12} y^{24}$
15. $x^{18} y^{30}$
17. $729y^8$
19. $144x^{14} y^{16}$

10.3
1. $14x^3 - 10x$
3. $-6a^4 + 12a^3 b$
   $-9a^2 b^2$
5. $8x^2 - 8x - 6$
7. $25x^2 - 1$
9. $49b^2 - 42b + 9$
11. $6x^2 - 2x - 28$
13. $21x^2 - 41x + 10$
15. $x^3 - x^2 - x + 10$
17. $y^3 - 5y^2 + 9y - 9$
19. $2x^4 - 12x^3 - x^2$
    $-x - 21$

10.4
1. 6
3. 2
5. 9
7. y^5
9. $x^2 y$

10.4 con't
11. $4x^3$
13. $4x(x - 2)$
15. $a^4(a - 3)$
17. $2x(2x^2 - x - 4)$
19. $3a(-a^2 - 2a + 4)$
    or $-3a(a^2 + 2a - 4)$

Chapter 10

Test A

1. $7x - 8$
2. $4x^2 + 2x$
3. $2x + 2$
4. $-3x^2 - 3x - 9$
5. 16
6. x^{13}
7. a^{35}
8. $27x^6$
9. $-6x^5$
10. b^{55}
11. $144a^{14}b^8c^{26}$
12. $15x^2 - 27x$
13. $-5x^3 + 35x^2 - 45x$
14. $x^2 - x - 20$
15. $9x^2 - 21x + 10$
16. $9x^2 - 25$
17. $16x^2 - 72x + 81$
18. $a^3 - 7a^2 + 3a + 27$
19. $A = \left(6x^2 - x - 2\right)$ in.
    $P = \left(16x - 4\right)$ in.
20. 18
21. $4y^2$
22. $5y(y - 3)$
23. $7a(2a - 1)$
24. $3\left(3x^2 - 2x - 4\right)$
25. $x^5\left(12x^2 - 11x + 7\right)$

Answers

Test B

1. b
2. c
3. a
4. c
5. d
6. a
7. b
8. c
9. d
10. b
11. c
12. d
13. a
14. b
15. d
16. c
17. d
18. a
19. b
20. c
21. b
22. a
23. d
24. c
25. a